Lecture Notes of the Institute for Computer Sciences, Social Informatics and Telecommunications Engineering 173

More information about this series at http://www.springer.com/series/8197

Jiafu Wan · Iztok Humar
Daqiang Zhang (Eds.)

Industrial IoT Technologies and Applications

International Conference, Industrial IoT 2016
Guangzhou, China, March 25–26, 2016
Revised Selected Papers

 Springer

Editors
Jiafu Wan
School of Mechanical and Automotive
  Engineering
South China University of Technology
Guangzhou
China

Daqiang Zhang
Software Engineering
Tongji University
Shanghai
China

Iztok Humar
Faculty of Electrical Engineering
University of Ljubljana
Ljubljana
Slovenia

ISSN 1867-8211   ISSN 1867-822X (electronic)
Lecture Notes of the Institute for Computer Sciences, Social Informatics
and Telecommunications Engineering
ISBN 978-3-319-44349-2   ISBN 978-3-319-44350-8 (eBook)
DOI 10.1007/978-3-319-44350-8

Library of Congress Control Number: 2016948761

Printed on acid-free paper

This Springer imprint is published by Springer Nature
The registered company is Springer International Publishing AG Switzerland

Preface

In recent years, the widespread deployment of wireless sensor networks, industrial clouds, industrial robots, embedded computing, and inexpensive sensors has facilitated industrial Internet-of-Things (IndustrialIoT) technologies and fostered some emerging applications (e.g., product lifecycle management). IndustrialIoT constitutes the direct motivation behind industrial upgrading (e.g., the implementation of smart factory of Industrie 4.0).

With the support of all kinds of emerging technologies, IndustrialIoT is capable of continuously capturing information from various sensors and intelligent units, securely forwarding all the data to industrial cloud centers, and seamlessly adjusting some important parameters via a closed loop system. Also, IndustrialIoT can effectively detect failures and trigger maintenance processes, autonomously reacting to unexpected changes in production. However, we are still facing some challenges, for example, it is very difficult to capture, semantically analyze, and employ data in a coherent manner from heterogeneous, sensor-enabled devices (e.g., industrial equipment, assembly lines, and transport trucks) owing to the lack of measurement tools, collection protocols, standardized APIs, and security guidelines.

2016 International Conference on Industrial IoT Technologies and Applications was held on March 24–26, 2016 in Guangzhou, China. The conference is organized by the EAI (European Alliance for Innovation). The Program Committee received over 60 submissions from 6 countries and each paper was reviewed by at least three expert reviewers. We chose 26 papers after intensive discussions held among the Program Committee members. We really appreciate the excellent reviews and lively discussions of the Program Committee members and external reviewers in the review process. This year we chose three prominent invited speakers, Prof. Min Chen; Prof. Lei Shu and Prof. Yan Zhang.

July 2016

Jiafu Wan
Iztok Humar
Daqiang Zhang

Conference Organization

Steering Committee Chair

Imrich Chlamtac   CREATE-NET and University of Trento, Italy

Steering Committee Members

Jiafu Wan   South China University of Technology, China
Min Chen   Huazhong University of Science and Technology, China
Daqiang Zhang   Tongji University, China

General Chair

Jiafu Wan   South China University of Technology, China

General Vice Chairs

Iztok Humar   University of Ljubljana, Slovenia
Daqiang Zhang   Tongji University, China

Technical Program Committee Co-chairs

Chin-Feng Lai   National Chung Cheng University, Taiwan
Jaime Lloret   Polytechnic University of Valencia, Spain
Tarik Taleb   Aalto University, Finland

Workshops Chair

Pan Deng   Chinese Academy of Sciences (ISCAS), China

Publicity and Social Media Chair

Houbing Song   West Virginia University, USA

Sponsorship and Exhibits Chair

Shiyong Wang   South China University of Technology, China

Publications Co-chairs

Hu Cai	Jiangxi University of Science and Technology, China
Zhaogang Shu	Fujian Agriculture and Forestry University, China
Jing Su	Guangdong Ocean University, China

Local Chair

Xiaomin Li	South China University of Technology, China
Hu Cai	JJiangxi University of Science and Technology, China

Web Chair

Hu Cai	Jiangxi University of Science and Technology, China

Technical Program Committee

Houbing Song	West Virginia University, USA
Li Qiu	Shenzhen University, China
Lei Shu	Guangdong University of Petrochemical Technology, China
Yan Zhang	Simula Research Laboratory, Norway
Yunsheng Wang	Kettering University, USA
Dewen Tang	University of South China, China
Yupeng Qiao	South China University of Technology, China
Leyi Shi	China University of Petroleum, China
Qi Jing	Peking University, China
Caifeng Zou	South China University of Technology, China
Seungmin Rho	Sungkyul University, South Korea
Pan Deng	Chinese Academy of Sciences (ISCAS), China
Kai Lin	Dalian University of Technology, China
Meikang Qiu	Pace University, USA
Feng Xia	Dalian University of Technology, China
Chi Harold Liu	Beijing Institute of Technology, China
Jianqi Liu	Guangdong University of Technology, China
Heng Zhang	Southwest University, China
Chao Yang	Institute of Software, Chinese Academy of Sciences, China
Tie Qiu	Dalian University of Technology, China
Guangjie Han	Hohai University, China
Feng Chen	Chinese Academy of Sciences, China
Dongyao Jia	University of Leeds, UK
Yin Zhang	Zhongnan University of Economics and Law, China
Qiang Liu	Guangdong University of Technology, China
Fangfang Liu	Chinese Academy of Sciences, China

Contents

IoT

Big Data

The Design and Implementation of Big Data Platform for Telecom Operators

Jing Tan[✉]

Shenyang Artillery Academy, Shenyang, China
tanj_tanjing@163.com

Abstract. This paper introduces the background of developing big data platform for telecom operators, and the benefits for telecom operators to using big data platform in some aspects that may improve forward and back changings. This document also presents a method of build big data platform using Hadoop according to the particularities of the data and systems in the telecom operators, and the implementation of this method in the internet department of a province-level telecom operator company.

Keywords: Big data · Telecom operators · Hadoop

1 Introduction

1.1 Background

Big data technology is broadly used in variety industries and companies providing technical support for marketing strategy. For instance, T-Mobile a Germany telecom operator use big data to integrate social media data, CRM and billing data that reduced customer churn rate into half in one season [1], and Walmart discern meaningful big data insights for the millions of customers to enjoy a personalized shopping experience with customers' shopping behavior data from on-and-off line [2].

Big data can be described by four characteristics [3] as follows:

(1) **Volume:** The size of data increases from TB to ZB with the growth of internet, mobile phone and sensors [4].
(2) **Variety:** Different with the structure data, types of unstructured data increases rapidly, like audio streams, video streams, images and geographic data.
(3) **Velocity:** The data is generated and processed fast to meet the demands and challenges of the companies' development and growth.
(4) **Veracity:** The quality of captured data can vary greatly. Accurate analysis depends on the veracity of source data.

Data generated from telecom operators also has these characteristics, take a middle class province of China Unicom as example, in 2012, internet access records reached 1 billion per day, and the quantity of these data is 9T per month [5]. Now, telecom operators start establishing big data platform and mining user profile to support business sales.

© ICST Institute for Computer Sciences, Social Informatics and Telecommunications Engineering 2016
J. Wan et al. (Eds.): Industrial IoT 2016, LNICST 173, pp. 3–11, 2016.
DOI: 10.1007/978-3-319-44350-8_1

1.2 Hadoop Introduction

Hadoop is a framework for distributed processing and Analysis of large data sets across clusters of computers using simple programing models [6]. Hadoop is originally present from Apache Nutch a sub-projects of Apache Lucene which is start from 2002 [7]. In 2004, Google published a paper entitled "MapReduce: Simplified Data Processing on Large Clusters" in OSDI (Operating System Design and Implementation) [8] which purpose MapReduce the most important modules in Hadoop for the first time. In the early 2008, Hadoop became Apache top-level project which including many sub-projects, such as Hive, HBase, Pig which are already graduated to be top-level projects in 2010. The core components of Hadoop framework consist of HDFS and MapReduce. HDFS (Hadoop Distributed File System) [9] is a distributed file system that provides high-throughput access to application data, in the meantime, MapReduce is a framework for job scheduling and cluster resource management system for parallel processing of large data sets.

Hadoop has five benefits [10] as follows:

(1) High reliability: Single-Point or Multi-Point failure cannot interrupt Hadoop's service.
(2) High scalability: Hadoop allocated and computed in the Hadoop cluster that could be easily scale to thousand node.
(3) High performance: Hadoop can move the data among the datanodes dramatically to guarantee the equilibrium of each nodes that would be fast in processing speed.
(4) High fault tolerance: Hadoop can automatically save the data in several copies, and can voluntarily relocate the jobs that are failure.
(5) Low-cost: Compare with database machine, business data warehouse and other data mart, Hadoop is an open source software that would substantially reduce the software cost in projects.

With the advantages of Hadoop, many companies chose Hadoop as framework to build up big data platform, including IBM, Adobe, LinkedIn, Facebook [11, 12], so Hadoop could be the choice for telecom operators.

2 The Importance of Developing Big Data Platform for Telecom Operators

Along with mobile network's development, the amount of mobile data increased a lot. However, the revenue of telecom operators does not increase as well. Moreover, the traditional income keep going down as it occupied by mobile data's generator (the third party business in substitution type). Even worse, Telecom operators is going to play as a channel. Therefore, how to take advantage of "channel" role, getting data resource from "channel", controlling another core-competitiveness outside of networking resource is the top question for Telecom operators, in developing mobile network business.

2.1 Improving Business Innovation Ability

Base on analysis of large amount of data, understanding customer's requirement, and then lead to business improvement. After business online, keep tracking and analyzing customer's behavior, such as how to find it, and ordering and usage, as well as any existing problems. These data is the foundation of making strategy for business improvement, enhance business's practicability and convenience, improving business quality and customer experience. Take network optimization as an example, we can use big data technique to analyze network traffic, and trend. Then modify resource configuration in short time, meanwhile, analyzing network log, improve the whole of network, and keep improve network quality and capacity, as well as customer's networking experience.

2.2 Improving the Efficiency of Marketing Promotion

Nowadays, Telecom operators still focus on fixed package in the aspect of traffic operation business, still using fixed pattern for setting package, instead of on user's demand. Base on analysis of user's requirement and characteristic of behavior, we can filter out the target user, matching right product, determining the good time for showing and selling for customer. Moreover, we can combine channel characters and channel execution, developing precision marketing that is based on requirement subdivision and users' precise positioning. Then enhance the standardization of customer's resource management, matching customer's requirement and product features, and finally raising customer's satisfaction and marketing efficiency.

2.3 Exploring New Business Mode

Exploring new business mode includes enhancement of traditional forward charging, as well as developing new mode of back charging.

(1) **Enhancement of Forward Charging:** By improving the ability of business innovation and smart marketing, Telecom operators' ability in forward charging will be improve. Base on that, Telecom operators is able to provide personalized service, and targeted products and services for different level users. Then raising product's value and enhance the ability of forward charging.

(2) **Exploring New Mode of Back Charging:** Refer to internet business mode; telecom operators could have variety of Back-charging business mode.

   (a) **Smart Marketing:** The profit model for Telecom operators is using big data technique to provide smart marketing and precise matching product requirement, combine with easy channel system. All of these can help business partner achieve sales targets rapidly, and then business partner will pay corresponding commission for Telecom operators.

   (b) **Consultant Services:** In the process of developing product, marketing planning and product optimization, Telecom operators provides comprehensive consultant service, which is based on data analysis for business partners, to improve the product's competitive and operating efficiency. Related consultant service is one of profit points for Telecom operators.

(c) **Precision Advertising:** Precision advertising is most valuable mode in back-charging model. Telecom operators has huge number of user groups that are all potential advertising audience. In the meanwhile, diversified media which telecom operators occupied covers multi-aspect of advertising audience become valid carrier of advertisement. More important is by controlling all aspect information of advertising audience; it is easy to achieve targeted advertising and effectiveness, and will be more attractive for advertisers.

2.4   Improving Influence of Industry Chain

Deep processing data, providing information service, and create more opportunity that is new for companies without violate user's privacy. Therefore, big data technology will help telecom operators transform from web service provider to information provider. The competition of mobile network is the competition of data scale and quality, instead of number of users, or product itself. The key action of improving influence of industry chain is trying to get more high quality data and controlling more key nodes of getting data.

3   The Big Data Platform Architecture for Telecom Operators

Big Data Platform Architecture: big data platform includes three main parts: Data Collection layer, Big Data layer and Data Sharing layer which is shown in Fig. 1. Data Sources provide the data that used for store and analysis.

3.1   Data Sources

Main data comes from three channels: user network accessing interface signaling data, internal system data, and internet spider data.

(1) **User network accessing interface signaling data** come from GB port, IUPS port, GN port, LTE port and WLAN port. All of these data is web pages' session via different network, including user-browsing website's IP address, time etc.
(2) **Internal system data** comes from internal operation system, such as BOSS (Business & Operation Support System), CRM (Customer Relationship Management), TAMS (Telecom Marketing & Analysis System). BOSS system consists of network management, system management, billing system, business, and finance, and customer service. BOSS system provides networking data, billing data, and customer data etc. CRM system is able to provide marketing data and user data. TMAS provides business data and customer consumption data etc.
(3) **Network spider data** mainly use spider to extract network information, and then provide data foundation that used to analysis customer's behavior of surfing on internet.

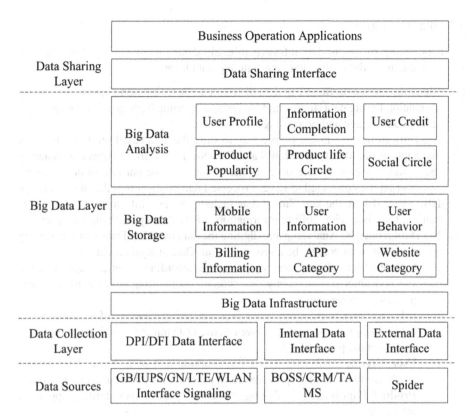

Fig. 1. The Architecture of Big data platform for telecom operators.

3.2 Data Collection Layer

Against three different data sources from Data Sources, use different way to collect data, include DPI/DFI data collection interface, internal data collection interface and external data collection interface.

(1) **DPI/DFI Data Collection Interface:** Collect user network interface signaling data, this system get the IP datagram from OBD (Optical Branching Device) which is connected in backbone network and analysis to the web session, so DPI data collection interface is handling user's web session records.

(2) **Internal Data Collection Interface:** Collect telecom operators' internal system data. This part already builds up relational database system, such as Oracle. Thus internal data collection interface is using JDBC or ODBC to get data from relational database system.

(3) **External Data Collection Interface:** Collect spider data. Network spider has sorted out the data that extract from internet, and store it in file system or database. Thus, external data collection interface is accessing file system or database system.

3.3 Big Data Layer

Big data layer provide big data infrastructure, big data storage and big data analysis, store the data from data collection layer into big data layer.

(1) **Big Data Infrastructure:** Big data distributed cluster base on Hadoop, providing foundation for big data storage and analysis, supporting high speed and high availability data storage and processing.

(2) **Big Data Storage:** Choose proper way to store the data based on data's features and application. Mainly use HDFS and HBASE. For DPI/DFI collected signaling data, original signaling data stores in file system. For the purpose of detail query, use external file connected to Hive. Processed daily data or monthly data stores in HBASE, IMEI are the key. Stored data includes Mobile Information, User Information, User Behavior, Billing Information, APP category, Website Category.

(3) **Big Data Analysis:** Analyzing and digging the data from Big Data Storage, getting new data that is supporting business extension. Data analysis includes:

  (a) **User Profile:** Labeling user's personality, according to gender, age, address, and consumer power, hobby etc., which is supporting smart marketing and precision advertising.

  (b) **Information Completion:** When user registering personal information, many data is incomplete or wrong, so we can use Data mining technique to complete or correct the personal data.

  (c) **User Credit:** Analyzing user's consumption and other basic data, getting user's credit evaluation, provides to Bank or Credit Information Company.

  (d) **Product Popularity:** Analyzing Telecoms' product popularity, supporting marketing.

  (e) **Product Life Circle:** Analyzing Telecoms' product life circle, understanding this product's operating.

  (f) **Social Circle:** Analyzing user's social circle, supporting smart marketing.

3.4 Data Sharing Layer

Data sharing layer adopt unified data accessing interface, open it to telecoms internal use, or open secure interface for external company to personal to use.

4   The Implementation of Big Data Platform

This architecture of big data platform for telecom operators is implemented in the internet department of a middle-class province for collecting and storing user internet accessing data, acquiring users' internet accessing behaviors. Combining with the users' demographic data, we use this platform to analysis the popularity and characteristic of the music, reading and game products of this department, and present marketing strategies for these products.

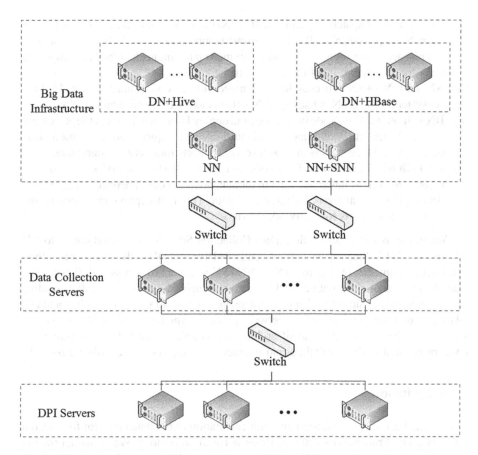

Fig. 2. The implementation of big data platform architecture in a province telecom operator.

Figure 2 shows the implementation of big data platform architecture in a province telecom operator. It includes two main parts, data collection servers and Big Data Infrastructure. In addition, DPI servers are very significant in telecom operators' network for gathering user internet accessing data but not a component in big data platform.

(1) **DPI Server:** DPI server is the data source of user internet data. DPI servers gather user signaling data from backbone network via OBD, and convert the user signaling data to user session data.

(2) **Data Collection Server:** It gets user session data from DPI server, and then sends to Big Data Infrastructure. Since the amount of session data is huge, it almost 300 thousand records per second, so this part consists of 6 servers, and each server need to process 50 thousand records per second on average.

(3) **Big Data Infrastructure:** This is a Hadoop cluster, which consists of servers including NN (NameNode) and DN (DataNode). NameNode is file system naming space in Hadoop that maintain the whole file system tree and all the related files and directories. DataNode is the file system's working node, it store and indexing

data base on dispatching with client or NameNode, also sending block list for NameNode periodically. Big Data Infrastructure is responsible for store the user Internet accessing session data and is summed up in minute, hour and day. Big Data Infrastructure is consist of three components:

(a) **NN and NN+SNN (Secondary NameNode):** NN is NameNode in Hadoop ecosystem. NN+SNN is backup of NN that is using HA, and used as SNN.
(b) **DN+Hive:** It is DataNode in Hadoop, and deploy Hive on it; it stores users' session data, and support session query, and multi-division's query and integration. This part deployed nine servers and each server supports four network interfaces.
(c) **DN+HBase:** Deploy both DataNode and HBase. It stores users' session details records, as well as integrated data in minute, hour and day. Therefore, it able to do detail query and analysis for integrated data. This part deploys nine servers and each server supports four network interfaces.

According to above description, Data Collection Servers can upload data into DN +Hive and DN+HBase concurrently. Each DN+Hive and DN+HBase server is setting two internet domains. One is for NN or NN+SNN servers to access and dispatch the data, the other is used for data collection servers to upload data to DN, so for avoiding confusion in the physic network and guaranteeing the upload speed, DN+Hive and DN +HBase is divided into two separate network segments which is connected by one switch separately. The fact is this kind of architecture is able to fulfill the requirement of concurrency that is about 300 thousand internet accessing session records per second.

5   Conclusion

This article introduced the design and implementation of big data platform for telecom operators. The Framework of this platform is based on Hadoop. We construct the big data platform that is special for Telecom operators. It is collecting users' internet accessing signaling data, internal system data and web spider data, improving the speeding of query data and data mining for users who are interested in music, reading and games, and providing guidelines for marketing strategy. As a result, this platform gain very good achievement.

Nowadays, telecom operator's traditional business, such as voice and SMS, this part's income is keep going down. Thus, Telecom operators is seeking new opportunities, many Telecom operators realized the value of big data, and already constructing big data platform. However, as the data is distributed in different BUs, it is very difficult to integrity data. In addition, different BU's business need is different, so it caused repeating construction of big data platform, and skill level also not the same. However, big data still quite important, it will bring more opportunities for Telecom Operators.

References

1. How to Use Big Data to Stop Customer. http://www.cio.com/article/2391831/big-data/how-to-use-big-data-to-stop-customer-churn.html
2. How Big Data Analysis helped increase Walmart's Sales turnover? http://www.dezyre.com/article/how-big-data-analysis-helped-increase-walmart-s-sales-turnover/109
3. Big data. https://en.wikipedia.org/wiki/Big_data#cite_note-INDIN2014-25
4. Segaran, T., Hammerbacher, J.: Beautiful Data: The Stories Behind Elegant Data Solutions. O'Reilly Media, Sebastopol (2009)
5. Wang, Z.J.: The application of hadoop in the telecom industry. Technical report, One China One World (2012)
6. Apache Hadoop. http://hadoop.apache.org/
7. Dean, J., Ghemawat S.: MapReduce: simplified data processing on large clusters. In: OSDI, vol. 51, no. 1, pp. 147–152 (2004)
8. Nagel, S.: Web crawling with Apache Nutch. In: ApacheCon EU 2014, Budapest (2014)
9. Shvachko, K., Kuang, H., Radia, S., Chansler, R.: The hadoop distributed file system. In: 2010 IEEE 26th Symposium Mass Storage Systems and Technologies, pp. 1–10. IEEE (2010)
10. Olson, M.: Hadoop: scalable, flexible data storage and analysis. IQT Q. 1, 14–18 (2010)
11. Borthakur, D., Gray, J., Sarma, J.S., Muthukkaruppan, K., Spiegelberg, N., Kuang, H., et al.: Apache hadoop goes realtime at Facebook. In: Proceedings of the 2011 ACM SIGMOD International Conference on Management of data. ACM (2010)
12. Sumbaly, R., Kreps, J., Shah, S.: The big data ecosystem at linkedIn. In: Ross, K.A., Srivastava, D., Papadias, D., (eds.) SIGMOD Conference, pp. 1125–1134. ACM (2013)

A Big Data Centric Integrated Framework and Typical System Configurations for Smart Factory

Shiyong Wang, Chunhua Zhang[✉], and Di Li

School of Mechanical and Automotive Engineering, South China University of Technology,
Guangzhou 510640, China
{mesywang,chhzhang,itdili}@scut.edu.cn

Abstract. Personalized consumption demand and global challenges such as energy shortage and population aging require flexible, efficient, and green production paradigm. Smart factory aims to address these issues by coupling emerging information technologies and artificial intelligence with shop-floor resources to implement cyber-physical production system. In this paper, we propose a cloud based and big data centric framework for smart factory. The big data on cloud not only enables transparency to supervisory control but also coordinates self-organization process of manufacturing resources to achieve both high flexibility and efficiency. Moreover, we summarize eight typical system configurations according to three key parameters. These configurations can serve different purposes, facilitating system analysis and design.

Keywords: Smart factory · Smart production · Smart product · Industry 4.0 · Industrial internet

1 Introduction

For a long time, shop-floor manufacturing resources in terms of machines and conveyers have been carefully organized to build production lines which are efficient and low-cost for mass production. However, the traditional production line is rather rigid so that it will lead to a long system down time and an expensive cost to change for another product type. To cope with ever increasing personalized consumption demands on multi-type and small- or medium-lot customized products, many advanced manufacturing schemes such as flexible manufacturing system (FMS) or intelligent manufacturing system (IMS) have been proposed. The researches on FMS expect to allocate manufacturing resources to a family of product types with a kind of central computerized controller [1, 2]. By contrast, the multi-agent system (MAS) method, a representative IMS scheme, models resources as autonomous agents that rely on peer to peer negotiation to dynamically reconfigure for different product types [3, 4].

Today, emerging information technologies raise credible opportunities to implement smart production. With cloud computing [5], big data [6–8], wireless sensor network (WSN) [9], Internet of Things (IoT) [10], and mobile Internet [11] et al. applied in manufacturing environment, machines, tools, materials, products, employees, and information systems (e.g., ERP and MES) can be interconnected and communicate with each other.

© ICST Institute for Computer Sciences, Social Informatics and Telecommunications Engineering 2016
J. Wan et al. (Eds.): Industrial IoT 2016, LNICST 173, pp. 12–23, 2016.
DOI: 10.1007/978-3-319-44350-8_2

This actually forms a manufacturing oriented cyber-physical system (CPS) [12, 13] or called cyber-physical production system (CPPS), which is the basis for smart factory termed by industry 4.0 initiative [14]. Compared with FMS and IMS, the smart production enabled by smart factory features high interconnection, mass data, and deep integration. Moreover, the product acts as a smart entity participating in the production process actively. Therefore, based on high bandwidth network and powerful cloud, smart production can implement high flexibility, high efficiency, and high transparency [15, 16].

In this paper, we propose a layered framework for smart factory to integrate shop-floor entities, cloud, client terminals, and people with industrial network and Internet. Big data and self-organization of smart shop-floor entities are two essential mechanisms to implement smart production. Big data enables transparency and coordinates self-origination process to achieve high efficiency. Self-organization makes reconfiguration process for multi-type products very flexible. To account for the diversity of shop-floor manufacturing resources, e.g., digital product memories (DPM) can be classified into storage, reference, autonomous or smart [17], and production execution can be alternative or hybrid, we propose an analysis model and identify three key parameters, according to which eight typical system configurations are recognized.

The article is organized as follows. In Sect. 2, we discuss the roles of cloud, big data, and self-organization in the smart production environment. In Sect. 3, the key parameters that affect reconfiguration ability, negotiation mechanism, and deadlock prevention are identified based on system analysis. In Sect. 4, eight typical configurations are constructed based on three key parameters, characteristics and application scenarios of which are further discussed. Finally, conclusions and future work are given in Sect. 5.

2 Integrated System Framework

Smart factory focuses on vertical integration of various components inside a factory boundary. It is a kind of manufacturing oriented cyber-physical system, i.e., cyber-physical production system that features high flexibility, high efficiency and high transparency. In this section, we present an integrated system framework and discuss related issues.

2.1 Cloud Based Integration

Figure 1 depicts a framework to integrate shop-floor entities, servers, client terminals, and people with industrial network and Internet. Shop-floor entities mainly include machines (for processing, assembling, testing, and storing et al.), conveyers (such as conveyor belts, AGVs, and loading/unloading robotic arms), and intelligent products (being processed by the system). Servers are specific computers for hosting various information systems such as ERP, MES, and CAX (CAD/CAE/CAM) systems. Client terminals in the form of computers and smart mobile phones et al. are for human-system interaction. People mainly refer to employees that distribute among various sectors, e.g., production, operation and maintenance, design, purchasing, sale, finance, and planning.

However, non-employees such as suppliers, customers, and supervisors can also link to this network.

Traditionally, separate servers are used for different information systems. However, with cloud computing technology, a network of servers can be virtualized as a huge resource pool to support elastic computing and storage demand. Therefore, different information systems can be deployed onto the single cloud platform, and distributed shop-floor entities and client terminals can be connected to the same cloud as well. As a result, all the enterprise activities ranging from design and production to management and planning are integrated based on cloud.

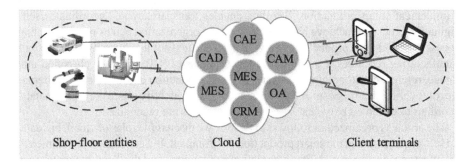

Fig. 1. Cloud based integrated framework of smart factory.

2.2   Big Data Based Fusion

Both shop-floor entities and client terminals can act as data terminals to gather various kinds of data to cloud (outer race of Fig. 2). However, the simple migration of information systems from separate servers to the single cloud is not enough to create meaningful big data. Today's information systems are designed to cope with different requirements, e.g., CAD for product definition, MES for production process management, and ERP for resource management. As a result, they will probably use different formats to describe the same data object causing inconsistency to block information flow among different systems.

For big data to come true, smart factory should be constructed in a data centric way (inner race of Fig. 2). A unified data model including vocabulary, syntax, and semantics should be defined to maintain consistency, continuity, and integrity of mass data. Therefore, different information systems can operate on the same data object set. As software modules interact with each other through data objects, tight logic coupling can be released so that information processing software can be further modularized and miniaturized (middle race of Fig. 2). This facilitates software deployment and lower cost, e.g., software modules can be selected on demand. Recall that both big data and information processing software run on cloud, whereas shop-floor entities and client terminals are connected with cloud through industry network or Internet.

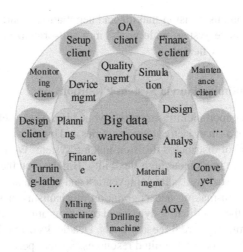

Fig. 2. Big data based fusion of smart factory.

2.3   Self-organization Based Resource Reconfiguration

The smart production system is designed for processing multiple types of products. The machines are redundant and the conveying system has multiple branches. Therefore, both machines and conveyers should be reconfigured dynamically. For example, one product may need machines 1, 3, and 5, whereas another product may need machines 2, 4, 6, as shown in Fig. 3. Obviously, products have to go through different branches to traverse the two different sets of machines. For distributed and autonomous machines and conveyers, negotiation based mechanisms are suitable for reconfiguration in a self-organized way.

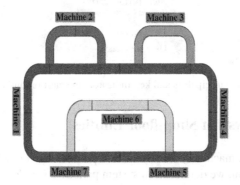

Fig. 3. Resource reconfiguration for different products.

In smart factory, the shop-floor entities are beyond the kind of numerical control systems that have abilities of computing, communication, control, sensing, and actuating. Smart entities can also make decisions by themselves and negotiate with others. Through autonomous decision-making and negotiation, smart entities cooperate with each other to achieve system-wide goals, in a self-organized way, making the reconfiguration process very flexible.

2.4 Performance Optimization via System Evolution

Cloud and network are important infrastructures, while big data and self-organization are essential mechanisms of a smart factory. Smart entities and big data analytics based coordinator construct a kind of distributed decision-making system. We rely on self-organization of smart entities to implement high flexibility. Big data, on the other hand, helps coordinate global efforts such as deadlock prevention and performance optimization.

When design decision-making and negotiation mechanisms of smart entities and customize behavior of the coordinator, dynamical reconfiguration, deadlock prevention, and performance optimization are three key goals. Deadlocks occur due to the fact that multiple products will compete for limited resources. System performance has a lot of indicators such as efficiency, utilization rate of machines, and load balance. Dynamical reconfiguration and deadlock prevention are fundamental requirements while system performance is desired to improve progressively with increasing experience and data. Moreover, user preferences can also affect system evolution, but they are generally preset and static. The related components and their relationship is shown in Fig. 4.

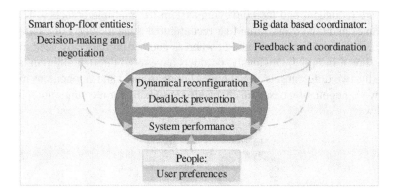

Fig. 4. Main participators and key indicators in smart production system.

3   System Analysis of Shop-floor Entities

As mentioned above, machines, conveyers, and products are main kinds of shop-floor entities. In this section, we define seven system parameters to describe system characteristics, three of which are further recognized as the key parameters.

3.1   System Analysis

Machines, conveyers, and products have some special characteristics which form a design space for constructing various production systems, as shown in Fig. 5.

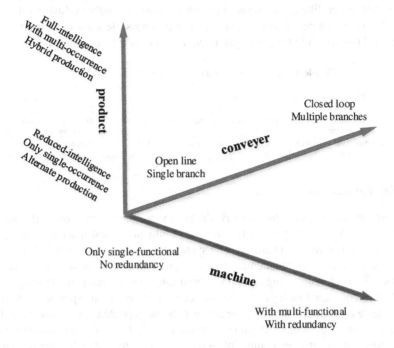

Fig. 5.   System configuration space in terms of machines, conveyers, and products.

For machines, one that has multiple sub functions is defined as multi-functional machine, whereas one that has only one function is defined as single-functional machine. If all the machines have different sub functions from each other, no functional redundancy exists, whereas two or more machines having the same sub functions introduces functional redundancy.

For conveyers, the resultant conveying route is either open or closed. Moreover, a route may have branches. While single open line is the simplest production line, multiple open lines can intersect with each other to form a complex route. Similarly, single loop is simple and applicable, whereas multiple loops can be linked together to build complex circular routes.

For products, the full-intelligence product can make decisions and negotiate with others by itself, whereas the reduced-intelligence product may do not have abilities of computing and communication, e.g., the product only attached with a RFID tag. Moreover, each product type specifies a sequence of operations. Therefore, in an operation sequence, if only one operation belongs to an operation type one defines the operation type as the single-occurrence (operation) type. If two or more operations belong to the same operation type, one defines the operation type as the multi-occurrence (operation) type.

The smart factory is for production of multiple types of products. However, this can be classified into alternate production (one type of products is processed after another) and hybrid production (multiple types of products are processed simultaneously).

The aforementioned characteristics are summarized as system parameters (Table 1) and each parameter, like Boolean variable, has only two mutually exclusive values. As the number of parameters is seven and each has two possible values, a variety of one hundred and twenty-eight different system configurations can be determined.

Table 1. System parameters and their allowed values.

		Parameters						
		1	2	3	4	5	6	7
Value	A	Only single-functional	No redundancy	Open line	Single branch	Reduced-intelligence	Only single-occurrence	Alternate production
	B	With multi-functional	With redundancy	Closed loop	Multiple branches	Full-intelligence	With multi-occurrence	Hybrid production

3.2 Key Parameters

For parameters 1, 2, 4, and 6, the value B addresses more general and practical situations than the value A does. For example, even one multi-functional machine can change parameter 1 from value A to value B. The system that allows functional redundancy is easier to deploy and the redundancy helps to guarantee robustness, e.g., in case of machine failure. The conveying system with multiple branches can extend to large space and adapt to complex topology. The multi-occurrence operation types are sometimes not avoidable considering the resource constraints and repeated operations. Therefore, when developing algorithms, the value for parameters 1, 2, 4, and 6 is assumed to be B; the developed algorithms are compatible with value A, as the value A addresses simple situations.

For parameter 3, the open lines are used widely in the traditional production lines. However, the open line will limit system's reconfiguration ability. As shown in Fig. 6, five machines for operation types A, B, C, D, and E are deployed along the unidirectional conveyor belt. Any product types that require operation sequences like [A, B, C], [B, C, D], and [A, C, E] can be processed as the fixed order from A to E is kept, whereas the operation sequences like [A, C, B] or [E, D, C] cannot be supported as the conveyor belt cannot route the products back from C to B or from E to D. By contrast, the closed loop conveying system like circular conveyor belt or bidirectional AGV can route products between any two machines. Therefore, this system parameter affects system reconfiguration ability, i.e., value B (closed-loop) can support complex reconfiguration whereas value A (open line) cannot.

For parameter 5, the product with full intelligence can participate in negotiation process as an active agent, whereas the product with reduced intelligence is passive and should rely on other components, i.e., machine or conveyer, to help it. Therefore, this parameter affects negotiation mechanism and negotiation process.

For parameter 7, the hybrid production is more complex than the alternate production. The hybrid production makes production process highly dynamical that deadlocks will occur unexpectedly. Therefore, the hybrid production needs more powerful

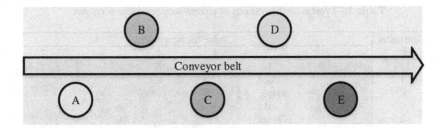

Fig. 6. Production system with open-line conveyor belt.

deadlock prevention strategy than the alternate production does. In a word, this parameter relates to deadlock prevention strategy.

In summary, the parameters 3, 5, and 7 are recognized as the key parameters. They respectively affect reconfiguration ability, negotiation mechanism, and deadlock prevention strategy, i.e., value A and B of these parameters require different strategies. As to parameters 1, 2, 4, and 6, the value B covers the application range of value A, so they are not treated as key parameters and only value B is considered during design.

4 Typical Configurations and Their Application

The three key parameters can be used to determine eight typical system configurations. Based on parameter 3, the eight configurations are divided into two groups. We formulate each configuration and discuss their distinct characteristics in this section.

4.1 Typical Configurations of Closed-loop Production System

Table 2 summarizes four typical configurations featuring closed-loop production system. The value of parameter 3 is B (closed loop) for these configurations, but the value combination of parameters 5 and 7 is different in each configuration.

Alternative Production VS Hybrid Production. As a general rule, efficiency increases with batch size. However, the alternative production is more sensitive to batch size than hybrid production, as illustrated in Fig. 7. This is because alternative production requires one type of products to be processed after another leading to system overhead in the case of product type switch. Hybrid production dose not suffer this kind of overhead, as it can accommodate multi-type products simultaneously. As a result, the hybrid production suits for small-lot production whereas the alternative production is more efficient for medium or mass production.

Full Intelligence VS Reduced Intelligence for Hybrid Production. Full intelligence product can carry and maintain its own data/state, and it can make decisions for itself. Therefore, full intelligence product is quit suitable to be used with hybrid production to maximize system performance. By contrast, reduced intelligence product will lower agility and efficiency when used with hybrid production, although it is cheaper. This is

Table 2. Typical configurations of closed-loop production system.

Configurat ion	Value for parameters						
	1	2	3	4	5	6	7
1	A /B	A /B	B (Closed loop)	A /B	A (Reduced-intelligence)	A /B	A (Alternate production)
2	A /B	A /B	B (Closed loop)	A /B	B (Full-intelligence)	A /B	A (Alternate production)
3	A /B	A /B	B (Closed loop)	A /B	B (Full-intelligence)	A /B	B (Hybrid production)
4	A /B	A /B	B (Closed loop)	A /B	A (Reduced-intelligence)	A /B	B (Hybrid production)

because reduced intelligence product needs to set up data structures for new types of products frequently in small-lot hybrid production.

Full Intelligence VS Reduced Intelligence for Alternative Production. Reduced intelligence product will not cause obvious performance loss and can save cost when it is used with medium or mass alternative production, as product type switch is not frequent in the case of large volume.

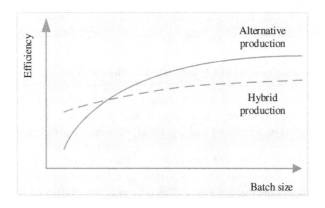

Fig. 7. Efficiency versus batch size for alternative and hybrid production.

In summary, the configuration 1 is quit suitable for medium or mass production, while the configuration 3 suits well for small-lot production. The configuration 2 can achieve equal efficiency as (or a little more than) configuration 1 but with much more cost. The configuration 4 cannot achieve equal efficiency as configuration 3 although it can save cost. These configurations enable users to balance between efficiency and cost based on batch size when design a smart factory.

4.2   Typical Configurations of Open-line Production System

Table 3 summarizes four typical configurations of open-line production system, where the value of parameter 3 is A (open line). Recall that the open line leads to very limited reconfiguration ability. The alternate production is possible as indicated in configurations 5 and 6, as long as the operation sequence is in accordance with the machine order. These two configurations suit for medium or mass production and the configuration 5 is cheaper than configuration 6. As to hybrid production, it is quite difficult to ensure the processing sequence of machines because of deadlock prevention, so configurations 7 and 8 are nearly not applicable.

Table 3. Typical configurations of open-line production system.

Configurat ion	Value for parameters						
	1	2	3	4	5	6	7
5	A /B	A /B	A (Open line)	A /B	A (Reduced-intelligence)	A /B	A (Alternate production)
6	A /B	A /B	A (Open line)	A /B	B (Full-intelligence)	A /B	A (Alternate production)
7	A /B	A /B	A (Open line)	A /B	B (Full-intelligence)	A /B	B (Hybrid production)
8	A /B	A /B	A (Open line)	A /B	A (Reduced-intelligence)	A /B	B (Hybrid production)

4.3   Algorithm Design for Typical Configurations

We have developed algorithms for configuration 2, and we find that negotiation process and deadlock prevention do not interrupt each other [18]. Therefore, if we could have developed algorithms for configuration 4, the resultant algorithms of configurations 2 and 4 can be used with configurations 1 and 3. Suppose that the negotiation mechanisms for reduced- and full-intelligence products are N1 and N2 respectively, and the deadlock prevention strategies for alternate and hybrid production are P1 and P2 respectively. Then the combination of N1 and P1 can be used to configuration 1, and the combination of N2 and P2 can be used to configuration 3. These strategies can also be used to configurations 5 to 8. However, special measures should be considered to account for the limited reconfiguration ability of open lines.

5   Conclusions and Future Work

By introducing cloud computing, big data, and artificial intelligence et al. into manufacturing environment, smart production is promising to achieve high flexibility, efficiency, and transparence. On one hand, smart shop-floor entities interact with each other to implement self-organization based dynamical reconfiguration. On the other hand, big data enables transparency for management and maintenance and coordinates system-wide goals such as deadlock prevention and performance optimization. The variety of physical shop-floor resources exist in the manufacturing environment, where machines, conveyers, and products are main participators. Many parameters relate to these resources and some of them play important roles in system design and analysis. The conveying route, product intelligence, production model are three key parameters to affect reconfiguration ability, negotiation mechanisms for dynamical reconfiguration, and strategies for deadlock prevention respectively. Based on these key parameters, we identify eight typical configurations, suitable for a range of applications. In the future, algorithms and practical experimental prototypes will be designed, implemented, and verified.

Acknowledgments. This work was supported in part by the National Key Technology R&D Program of China under Grant no. 2015BAF20B01, the Fundamental Research Funds for the Central Universities under Grant no. 2014ZM0014 and 2014ZM0017, he Science and Technology Planning Project of Guangdong Province under Grant no. 2013B011302016 and 2014A050503009, and Science and Technology Planning Project of Guangzhou City under Grant no. 201508030007.

References

1. Balogun, O.O., Popplewell, K.: Towards the integration of flexible manufacturing system scheduling. Int. J. Prod. Res. **37**(15), 3399–3428 (1999)
2. Priore, P., de la Fuente, D., Puente, J., Parreño, J.: A comparison of machine-learning algorithms for dynamic scheduling of flexible manufacturing systems. Eng. Appl. Artif. Intell. **19**(3), 247–255 (2006)
3. Leitão, P.: Agent-based distributed manufacturing control: a state-of-the-art survey. Eng. Appl. Artif. Intell. **22**(7), 979–991 (2009)
4. Shen, W., Hao, Q., Yoon, H.J., Norrie, D.H.: Applications of agent-based systems in intelligent manufacturing: an updated review. Adv. Eng. Inform. **20**(4), 415–431 (2006)
5. Xu, X.: From cloud computing to cloud manufacturing. Robot. Comput. Integr. Manufact. **28**(1), 75–86 (2012)
6. Liu, Q., Wan, J., Zhou, K.: Cloud manufacturing service system for industrial-cluster-oriented application. J. Internet Technol. **15**(4), 373–380 (2014)
7. Chen, M., Mao, S., Liu, Y.: Big data: a survey. Mobile Netw. Appl. **19**(2), 171–209 (2014)
8. Chen, F., Deng, P., Wan, J., Zhang, D., Vasilakos, A., Rong, X.: Data mining for the internet of things: literature review and challenges. Int. J. of Distrib. Sens. Netw. **2015**, 1–12 (2015)
9. Qiu, M.K., Xue, C., Shao, Z., Zhuge, Q., Liu, M., Sha, E.H.-M.: Efficent algorithm of energy minimization for heterogeneous wireless sensor network. In: Sha, E., Han, S.-K., Xu, C.-Z., Kim, M.-H., Yang, L.T., Xiao, B. (eds.) EUC 2006. LNCS, vol. 4096, pp. 25–34. Springer, Heidelberg (2006)
10. Tao, F., Zuo, Y., Xu, L.D., Zhang, L.: IoT based intelligent perception and access of manufacturing resource towards cloud manufacturing. IEEE Trans. Ind. Inform. **10**(2), 1547–1557 (2014)
11. Frazzon, E.M., Hartmann, J., Makuschewitz, T., Scholz-Reiter, B.: Towards socio-cyber-physical systems in production networks. Procedia CIRP **7**, 49–54 (2013)

12. Riedl, M., Zipper, H., Meier, M., Diedrich, C.: Cyber-physical systems alter automation architectures. Ann. Rev. Control **38**(1), 123–133 (2014)

13. Wan, J., Zhang, D., Sun, Y., Lin, K., Zou, C., Cai, H.: VCMIA: a novel architecture for integrating vehicular cyber-physical systems and mobile cloud computing. Mobile Netw. Appl. **19**(2), 153–160 (2014)

14. Recommendations for implementing the strategic initiative INDUSTRIE 4.0. http://www.acatech.de/ fileadmin/user_upload/Baumstruktur_nach_Website/Acatech/root/de/Material_fuer_Sonderseiten/ Industrie_4.0/Final_report__Industrie_4.0_accessible.pdf

15. Wang, S., Wan, J., Li, D., Zhang, C.: Implementing smart factory of Industrie 4.0: an outlook. Int. J. Distrib. Sens. Netw. (2015, in press)

16. Wan, J., Yan, H., Liu, Q., Zhou, K., Lu, R., Li, D.: Enabling cyber-physical systems with machine-to-machine technologies. Int. J. Ad Hoc Ubiquitous Comput. **13**(3/4), 187–196 (2013)

17. Herzog, G., Kröner, A.: Towards an integrated framework for semantic product memories. In: Wahlster, W. (ed.) SemProM, pp. 39–55. Springer, Heidelberg (2013)

18. Wang, S., Wan, J., Zhang, C., Li, D.: Towards smart factory for industry 4.0: a self-organized multi-agent system with big data based feedback and coordination. Comput. Netw. (2015, in press)

Data Acquisition and Analysis from Equipment to Mobile Terminal in Industrial Internet of Things

Minglun Yi[1], Yingying Wang[2], Hehua Yan[2(✉)], and Jiafu Wan[3]

[1] Jiangxi University of Science and Technology, Jiangxi, China
15692402321@163.com
[2] Guangdong Mechanical and Electrical College, Guangzhou, China
wyybaby@163.com, hehua_yan@126.com
[3] South China University of Technology, Guangzhou, China
Jiafuwan_76@163.com

Abstract. Internet of things (IoT) has become more and more popular in the new information technology era. It is also an important stage that supplies development of the information age. The aim of the IoT is to shorten the distance between objects and objects through the information system. This paper constructs an application model of the IoT and mobile terminal to realized human-machine interaction management, based on Android APP and ThingSpeak cloud platform to achieve the data signals' communication between intelligent mobile terminal and the different environments of device at anytime or anywhere. The model with Arduino development board as the underlying controller by the ESP8266 serial wireless WiFi module is linked into the Internet. In this way, the acquisition signal from the bottom control terminal will be sent to the cloud platform. Writing control program for the mobile terminal and the collection of real-time temperature or humidity parameter information, through be linked into 3 G/4 G network or WiFi router network to access cloud platform for data query and monitoring equipment.

Keywords: Android · Cloud platform · Mobile terminal · IoT

1 Introduction

The rapid development of Internet of Things (IoT) is considered to be a significant progress and opportunity in the field of information technology. IoT aims at assisting human to realize human-computer interaction and artificial intelligent [1].

Reference [2] summed up two major themes in Industries 4.0: smart factory and intelligent production that is a group consisting of machines will be self-organize, and the supply chain will be automatically coordinated. However, Ref. [3] pointed out that the current theoretical research of IoT is still in development stage, there some network should be accurately called Intranet of things at present, which be used to link to objects without the ubiquitous connectivity of internet. Recently, cloud computing is an emerging technology for improving inter-connectiveness of things via assistance of clouds. The platform of cloud will be a critical factor in the intelligentization of IoT.

© ICST Institute for Computer Sciences, Social Informatics and Telecommunications Engineering 2016
J. Wan et al. (Eds.): Industrial IoT 2016, LNICST 173, pp. 24–35, 2016.
DOI: 10.1007/978-3-319-44350-8_3

Based on the above research, in this paper we constructed a model, which adopt the cloud platform as hub center for information interchange, and the cell phone as mobile terminal to access cloud platform to simulate the mobile IoT which has the characteristic of the Internet, so as to reflect the application of a feature of industrial 4.0.

The rest of the paper is organized as follows: In Sect. 2, mainly introducing the application and research of cloud computing and mobile terminal in the field of IoT in recent years. The overall architecture and enabling technology for mobile IoT model is introduced in Sect. 3. Following that, the design method and phase of mobile IoT model is described in Sect. 4. In Sect. 5, the result of experiment that this model is presented. Finally, Sect. 6 concludes the paper.

2   Related Works

2.1   Related Research on Cloud Computing in the IoT

The IOT as a typical information and communication system, not only existing the ability of the Internet to store and transmit the information, but also can automatically collect and process the information of things. Therefore, the IoT must have the functional characteristics of the Internet for constructing the global information infrastructure to link to objects [3]. With the development of cloud computing, cloud platform [4] can provide an excellent environment for massive data qualitative analysis processing, and form a visual, intuitive and provide decision reference data set [5].

Cloud computing is a computing mode based on the Internet. By this way, the sharing of software/hardware resources and information can be available to the computers and other equipment on demand. Therefore, the IoT' structure and application will have big development based on the technology introduction of cloud computing. Reference [6] pointed out the advantages of the function and flexibility by the means of the smart home based on cloud computing compared with the traditional home automation system. The intelligent community management and control system based on cloud computing is introduced in the [7]. Reference [8] proposed that the system based on cloud computing system for realizing the intelligent community management and control; the authors design the manufacturing service model based on cloud; Reference [9] proposed sensor cloud concept and technology; Reference [10] designed the cloud services of the reservoir scheduling automation system transformation and upgrading, discusses the cloud computing in the Internet of things as a comprehensive information processing platform. Therefore, comparing to the traditional IoT, cloud computing makes IoT more intelligent, scalable, stability, but also make its application management more transparent and convenient.

2.2   Related Works on the Mobile Terminal in IoT

Development of information technology is to using the Internet application as the basement to realize the interaction of terminal equipment [11]. With the rapid development of science and technology, the intelligent mobile terminal change to a comprehensive information processing platform from a simple communication tools [12]. Just as the mobile terminal equipment with the functions of data collection, processing and transmission, etc., so the development of mobile terminal equipment will enable to achieve the interaction between people and things in IoT. Such as Ref. [13] exhibited that the design of network remote monitoring and control system for aquaculture Android platform based on Internet; And the application of the mobile terminal in the logistics information system based on IoT is described in Ref. [14]; Reference [15] discussed that the mobile terminal in the IOT application role; Reference [16] pointed out that the needs of development of mobile terminal in environment of IoT, etc., through these application forms of mobile terminal in IoT, these papers discusses it as a communication medium or carrier between people and things for the information exchange.

Therefore, with the development of information communication technologies, IoT mobile terminals will become more and more popular in industrial field.

3 Key Enabling Technology

This paper according to the model of "bottom controller—cloud platform—mobile terminal" to research and discuss the relevant techniques of IoT information interaction [17], and to simulate a data exchange format from Fig. 1, namely the process of from data acquisition to mobile terminal data displays: sensor—Arduino development board + ESP8266 WiFi module—ThingSpeak cloud platform—smart phone. The technology and facilities involved in this process are: network resource access technology, ThingSpeak cloud platform, WiFi module mode setting and mobile terminal development system selection.

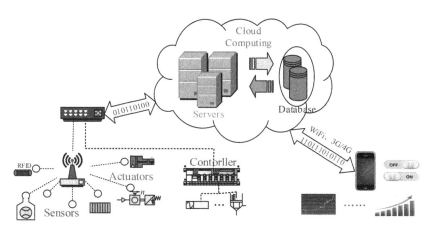

Fig. 1. Application model diagram

3.1 Network Resource Access Technology

In the IoT, each kind of resource is relatively independent, with independent access address and life cycle. This paper considered heterogeneous resources interoperability problems are caused by the information of different devices producing, processing and receiving [18], therefore, combining with the conclusion of literature [19], introducing the semantic web technology to the model of information interaction among. Each of the "resources" generated by this model is an addressable entity, the Uniform Resource Locator (URL) provide an abstract identification method to the location of the resources, so used this method to determine the location of the resources. Therefore, between the application-systems can rely on the resources location method for data transmission.

There are three elements for data transmission between the application-systems: transmission mode, transmission protocol, data format.

3.1.1   Transmission Mode

Transmission methods use the Socket method, it is the simplest way of interaction, and is a typical C/S interaction mode. The client terminal connects to the server through the IP address and port designated for the message exchange.

In this paper, the bottom controller timing sampling data and on time to upload, so according to the equipment type and the real time of the data transmission by socket method makes the underlying control terminal through the WiFi module to connect to the cloud server platform.

3.1.2   Transport Protocol

Transmission protocol, this paper considers the use of TCP/IP protocol and Http protocol. TPC protocol is a transport layer connection oriented, and end-and-end data packet transmission protocol. It's mainly used for solution how to data transmission in the network. Comparing with the non-connected oriented of UDP protocol, the TCP protocol transmission data reliability is higher. However, HTTP is the application layer protocol, mainly to solve how to pack data, mobile terminal applications program can be achieved access to cloud platform resource interface through the use of POST, PUT, GET, DELETE operations of HTTP.

3.1.3   Data Format

This paper adopts the JavaScript Object Notation (JSON) data format for retrieving data from web services. JSON which is a lightweight data interchange format based on the JavaScript programming language, data format is relatively simple, and easy to read/write. JSON is mainly used for with server data exchange, due to its format is compressed and occupy bandwidth is small, and easy to parse, and support multiple languages.

3.2   ThingSpeak Cloud Platform

Cloud platform in the field of IoT as a network hub center is used for the exchange of information [20]. Cloud platform provide the API address to share software and hard-ware resources and information to other devices. Therefore, this paper uses ThingSpeak cloud platform to carry on the experiment.

ThingSpeak cloud platform is an open source cloud IoT platform for constructing IoT applications and provides specialized services for user, the user can create multiple channels in this platform, and each channel provides eight fields to the same terminal for acquisition of eight different data. ThingSpeak can handle HTTP requests, and store and process data. The key features of this open data platform include: open API, real time data collection, location data, data processing and visualization, device status messages and plug-ins.

3.3 WiFi Module Mode Settings

ESP8266 WiFi module has two patterns: AP pattern and the STA pattern [21], the AP pattern is the wireless access point pattern, the WiFi model is a creator of wireless network, also is the center node of the network, under this pattern. General office and home use wireless router as an AP. And STA pattern for the site pattern, STA is refers to each terminal which is connection to the wireless network (such as notebook computers, PDA and other networking terminal) can be called a site.

In this paper, the model of experiment is use of WiFi ESP8266 module to connect the wireless router, and the data which is collected by Arduino development board will via the Internet to upload to ThingSpeak cloud platform. Therefore, in the needs of experiment, the WiFi ESP8266 module is set to STA mode.

3.4 Mobile Terminal Development System Selection

Using mobile terminal access to the IoT to achieve mobile Internet, it is necessary to carry on personalized mobile application software development. Currently, the system for mobile application software development is mainly divided into iOS and Android in the market.

The iOS originated in the Apple Corp OSX, it's based on the UNIX system. The iOS and the revaluate equipment are closely integrated, the current point of view that the integration of the iOS device and drive optimization comparing with similar products is the most outstanding. But the drawback of IOS system is controlled strictly by apple. In most cases, the other party application is unable to get the iOS' entire API, and its development environment must be the Mac operating system, developed application cannot be applied to the other products equipment. Therefore, using iOS in this paper' development conditions of mobile terminal model design are relatively harsh.

Android is a open source operating system base on the Linux and JAVA, although its performance is not as flexible and stable iOS, but this difference will be more and more small with the improving of Google. Due to the openness of the Android platform, it can do much more than iOS. Therefore, this model mobile terminal development using Android phone as a client terminal, and using Eclipse IDE to develop the android app which with links of ThingSpeak cloud platform corresponding API addresses, and download to the cell phone, through the friendly user interface, and guide the user to select the interface of corresponding parameters for related data query [22].

4 Design Methods

In this paper, the experimental model design is divided into three parts: (1) Arduino development broad communicates with cloud platform, (2) Android mobile phone communicates with cloud platform, and (3) Data display. Through these 4 designs, to describe the experimental model as an IoT model with function of the Internet, this can be used to reflect the function features of this model.

4.1 Communication Between Arduino Development Board and Cloud Platform

Using the Arduino development board to upload sampling data to ThingSpeak cloud platform through the ESP8266 WiFi module, this process need to meet two conditions: 1. Calling the ThingSpeak data channel address API; 2. ESP8266 WiFi module serial mode setting.

4.1.1 Calling the ThingSpeak Data Channel Address API

The Arduino IDE software programming need to call data channel write-API addresses of ThingSpeak cloud platform (GET/update? api_key = Write_API_KEY STRING & FIELD_NAME = VALUE), in the program by write data to the VALUE, and combined with the API send the VALUE to the ThingSpeak server corresponding storage area.

4.1.2 ESP8266 WiFi Module Serial Mode Settings

Due to this IoT model need to communicate with cloud platform, therefore, the WiFi module needs to be set to STA pattern by connecting the wireless router, and then access to the Internet for data communicate with cloud platform. Due to ESP8266 WiFi module access object is the cloud server, so the port type is the client and the module adopt the TCP transmission [23]. In this paper, experiment with in ThingSpeak cloud platform for model design, so the remote server IP address settings is 184.106.153.149 or api.thing-speak.com,and the port number of the remote server is 80, in the WiFi module.

4.2 Communications Between Android Mobile Phone and Cloud Platform

When using a cell phone to query the data which collected by ThingSpeak cloud platform, it is required to use the Http protocol to access the corresponding website of ThingSpeak to send GET request to get the data. Because of the user terminal access network is a time-consuming process, in order to prevent the UI thread is blocked lose response to user actions, Android provides an abstract class AsyncTask<Parma, progress, Result>, and makes network access process can be a simple asynchronous processing. Therefore, in terms of network communications programming by using of inheriting the AsyncTask class to handle issues that the user's cell phone access Thing-Speak cloud platform.

Due to ThingSpeak cloud platform have simply store and analysis for these sampling information (for example, averaging or accumulating the sampling data at a certain period of time, etc.), therefore, users only need to call its API addresses that can be queried the corresponding results. Such as the API address of the real time data acquisition (https:// api.thingspeak.com /channels/ ThingSpeak_CHANNEL_ID/feeds /last?api_key=Read API_KEY STRING) and the API address of the historical data aquisition (http://api.thing-speak.com/channels/ThingSpeak_CHANNEL_ID /feeds.json? api_key=Read API_KEY STRING&average= T &start= Ts UTC &end= Te UTC), and then read the data to JSON data analysis [24].

4.3   Data Display

Data display need to develop an android APP software, the software is used for the user can access information in the database of server via the Internet using mobile communications tool. The software interfaces consist of a main interface and four graphical display interfaces. The jump between the interfaces is performed by triggering the event of the corresponding button. As shown in Fig. 2.

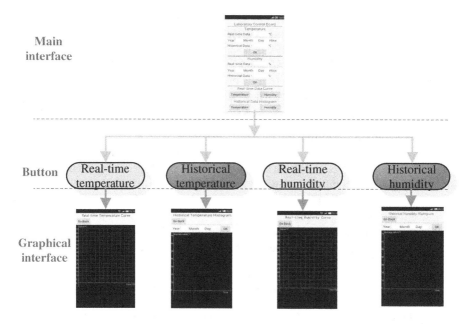

Fig. 2. Overall interface design

The design of data display program is divided into two mainly parts: 1. Real-time data display; 2. Historical data query & display.

4.3.1   Real Time Data Display

The Real time data display that the cell phone per 10 s sends instructions to ThingSpeak server to obtain real time update of data for digital display or graph shows. The real-time data acquisition program has mainly three threads, namely the UI thread and timer timing processing thread and network communication thread.

The UI thread is mainly responsible for handling user event which is user' operation and interface display, and so on. Considering the Android UI thread is not safe, which means that if there are multiple threads to manipulate a UI component, it may lead to thread safety issues. Therefore, in this program adopt the Handler message transfer mechanism, using a timer timing to send a message to a message-queue and Handler from the message-queue gets the message, and triggering Asynctask' subclass for the network data acquisition, so as to achieve the purpose of real-time data updating. Details of the process are shown in Fig. 3.

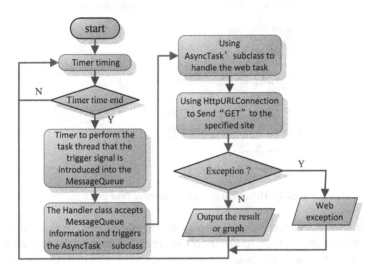

Fig. 3. Real time data display program flow chart

4.3.2   Historical Data Query and Display

Historical data query & display that according to the information of time setting (the time must be in the data sampling of the total time range) send instructions to the Thing-Speak server to gets that it in a set time period the average value of sampling data for digital display or bar graph display.

Historical data acquisitions program mainly rely on the user to the cell phone screen text edit box enter the value of query time, and by clicking the query button to trigger the AsyncTask' subclass for network data acquisition. With the cell phone APP program in the history of the temperature value query as an example, ThingSpeak cloud platform timing for data collection, and provides the different API key of data processing. Therefore, when connected to the platform by the use of HttpURLConnection class, according to the characteristics in the use of the API key, to obtain specific data value, such as a certain period of time the average value, maximum value. Details of the process are shown in Fig. 4.

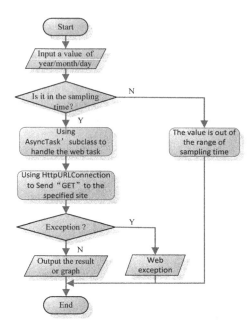

Fig. 4. Historical data query display flow chart

Through these above methods can display digital information, dynamic data graph and image information for user, so that users can clearly understand the operation of the things that they are concerned about.

5   Experimental Results

Based on temperature query as an example, the processing results of cell phone app as shown in Fig. 5, figure (a) is the main interface of numerical display, figure (b) is real-time temperature curve, and figure (c) is historical temperature histogram.

In the model designed in this paper, the cell phone can use WiFi or 3 G/4 G signals for information query at anytime or anywhere, cloud platform can be applied to the data processing and transmission of multiple models, the bottom controller can be flexibly using wireless technology for information exchange. By analogy, this model can be added more new functions to be improved, through the data mining technology in the IoT generated huge amounts of data to extract hidden information, analysis and modeling, and be applied to industrial production management [25]. The mobile terminal' design of interface program is on-demand, and the cloud platform by industrial production model can be flexibility customized, and the bottom control equipment according to the function module to be distinguished, these flows of information are gathered to upload to the cloud by using the wireless communication technology, combined with the industrial model setting and enterprise management personnel adjustment measures, so the cloud platform will automatic provide feedback information and assign production task to other equipments in factory [26, 27], that to achieve the wisdom of plants, intelligent production and the mobile Internet.

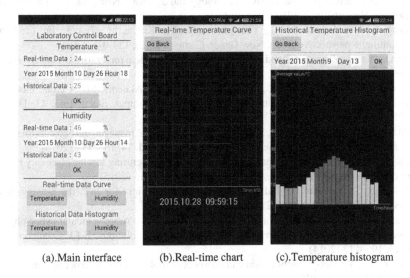

(a).Main interface   (b).Real-time chart   (c).Temperature histogram

Fig. 5.   Data processing results

6   Conclusions

Based on cloud platform and intelligent cell phone, this paper realized the mobile Internet with the Internet function: a model of the interaction between human and things information. However, to realize the automation and intelligence of the true meaning of IoT, still need further theoretical innovation and research, searching for common characteristics from the various fields of IoT applications, refining cloud computing application category, so that it can be used to flexibly make decisions and offer intensive application platform for industries in the fields of management, production sales, and so on.

Acknowledgment.   This work is partially supported by the National Natural Science Foundations of China (Nos. 61572220, and 61262013), the Fundamental Research Funds for the Central Universities (No. 2015ZZ079), the Natural Science Foundation of Guangdong Province, China (No.2015A030313746) and the Science and Technology Planning Project of Guangdong Province, China (No. 2013B011302016).

References

1. Wan, J., Yan, H., Liu, Q., Zhou, K., Lu, R., Li, D.: Enabling cyber-physical systems with machine-to-machine technologies. Int. J. Ad Hoc Ubiquitous Comput. **13**(3/4), 187–196 (2013)
2. Wan, J., Cai, H., Zhou, K.: Industrie 4.0: enabling technologies. In: IEEE 2014 International Conference on Intelligent Computing and Internet of Things, pp. 135–140 (2015)

3. Zou, C., Wan, J., Chen, M., Li, D.: Simulation modeling of cyber-physical systems exemplified by unmanned vehicles with WSNs navigation. In: Park, J.J(.J.H.)., Jeong, Y.-S., Park, S.O., Chen, H.-C. (eds.) EMC Technology and Service. LNEE, vol. 181, pp. 269–275. Springer, Heidelberg (2012)
4. Wang, L., Von Laszewski, G., Younge, A., He, X., Kunze, M., Tao, J., Fu, C.: Cloud computing: a perspective study. New Gener. Comput. 28(2), 137–146 (2010)
5. Liu, Q., Wan, J., Zhou, K.: Cloud manufacturing service system for industrial-cluster-oriented application. J. Internet Technol. 15(3), 373–380 (2014)
6. Korkmaz, I., Metin, S., Gurek, A., Gur, C., Gurakin, C., Akdeniz, M.: A cloud based and Android supported scalable home automation system. Comput. Electr. Eng. 43, 112–128 (2014)
7. Mital, M., Pani, A., Damodaran, S., Ramesh, R.: Cloud based management and control system for smart communities: a practical case study. Comput. Ind. 74, 162–172 (2015)
8. Guo, L., Wang, S., Kang, L., Cao, Y.: Agent-based manufacturing service discovery method for cloud manufacturing. Int. J. Adv. Manuf. Technol., 1–15 (2015)
9. Distefano, S., Merlino, G., Puliafito, A.: A utility paradigm for IoT: the sensing cloud. Pervasive Mob. Comput. 20, 127–144 (2014)
10. Wan, J., Zhang, D., Zhao, S., Yang, L.T., Lloret, J.: Context-aware vehicular cyber-physical systems with cloud support: architecture, challenges and solutions. IEEE Commun. Mag. 52(8), 106–113 (2014)
11. Zhu, Z., Lu, Y.: Hotspot application and prospect of mobile phone Internet of things. China Internet 11, 48–50 (2010)
12. Patel, M., Nagl, S.: Discussion. In: Patel, M., Nagl, S. (eds.) The Role of Model Integration in Complex Systems Modelling. UCS, vol. 6, pp. 127–152. Springer, Heidelberg (2010)
13. Li, H., Liu, X., Huan, J.: Remote monitoring and control system for aquaculture based on IoT and Android platform, CN203299614U (2013)
14. Du, X., Tang, B., Wu, F.: Design and implementation of IoT logistics management system based on Android terminal. Comput. Eng. Softw. 12, 26–31 (2013)
15. Ma, J.: Mobile terminal rich Internet applications. World Telecommun. 7 (2011)
16. Li, J.: Thinking on the development trend of mobile terminals in the Internet of things. In: Beijing Youth Communication Science and Technology Forum 2010 (2010)
17. Jing, Q., Vasilakos, A., Wan, J., Lu, J., Qiu, D.: Security of the Internet of things: perspectives and challenges. Wireless Netw. 20(8), 2481–2501 (2014)
18. Liu, J., Wang, Q., Wan, J., Xiong, J.: Towards real-time indoor localization in wireless sensor networks. In: Proceedings of the 12th IEEE International Conference on Computer and Information Technology, Chengdu, China, pp. 877–884, October 2012
19. Lin, Y.: Research on the key technology of semantic Web service composition and its application in Internet of things. South China University of Technology (2013)
20. Shu, Z., Wan, J., Zhang, D., Li, D.: Cloud-integrated cyber-physical systems for complex industrial applications. Mob. Netw. Appl. (2015). doi:10.1007/s11036-015-0664-6. ACM/Springer
21. Friedman, R., Kogan, A., Krivolapov, Y.: On power and throughput tradeoffs of WiFi and bluetooth in smartphones. IEEE Trans. Mob. Comput. 12(7), 1363–1376 (2013)
22. Aanensen, D., Huntley, D., Feil, E., al-Own, F., Spratt, B.: EpiCollect: linking smartphones to web applications for epidemiology, ecology and community data collection. PloS one 4(9) (2009)
23. Kim, T., Lee, J., Cha, H., Ha, R.: An energy-aware transmission mechanism for WiFi-based mobile devices handling upload TCP traffic. Int. J. Commun Syst 22(5), 625–640 (2009)

24. Merelo-Guervós, J., Castillo, P., Laredo, J., Prieto, A.: Asynchronous distributed genetic algorithms with Javascript and JSON. In: IEEE World Congress on Computational Intelligence in Evolutionary Computation, pp. 1372–1379 (2008)
25. Chen, F., Deng, P., Wan, J., Zhang, D., Vasilakos, A., Rong, X.: Data mining for the Internet of things: literature review and challenges. Int. J. Distrib. Sens. Netw. **2015**(431047), 14 (2015). doi:10.1155/2015/431047
26. Wang, S., Wan, J., Zhang, D., Li, D., Zhang, C.: Towards the smart factory for industrie 4.0: a self-organized multi-agent system assisted with big data based feedback and coordination. Elsevier Comput. Netw. (2016). doi:10.1016/j.comnet.2015.12.017
27. Wang, S., Wan, J., Li, D., Zhang, C.: Implementing smart factory of industrie 4.0: an outlook. Int. J. Distrib. Sens. Netw. **2016**, Article ID 3159805, 10 pages (2016). doi:10.1155/2016/3159805

Research About Big Data Platform of Electrical Power System

Dongmei Liu, Guomin Li, Ruixiang Fan, and Guang Guo[✉]

State Grid Information and Telecommunication Branch, Beijing 100761, China
43383047@qq.com

Abstract. Along with the construction of intelligent power grid and the continuous expansion of it, power systems produce large amounts of data, it is particularly important to integrate, analysis and process, and traditional data processing technology is difficult to meet the demand. The technology of big data injects new vitality to the development of intelligent power system, in the area of power system mastering the key technology of big data for the sustainable development of the electric power industry and the establishment of a strong smart grid is of great significance. Firstly, this paper does research about the key technology and processing scheme due to big data in power system. Secondly, it explores big data in power system based on cloud computing architecture.

Key words: Power system · Big data · Cloud computing

1 Introduction

Although the industry has some consensus, the definition of big data did not have a unified definition. McKinsey suggested "big data refers to the size beyond the typical data capture, storage, management and analysis scope of data collection"; Gartner said "big data need to deal with the new model that can have better decision-making, insight found mass force and the process optimization ability, high growth and diversification of information assets". In 2011, the Science pushed out the special issue name "Dealing With Data", deeply discussed challenges the Data deluge (DD) bring about, and pointed out that people would get more opportunities to play a great role of science and technology to promote social development if these huge amounts of Data could be more effectively organized and used [1].

As the energy supply system, economic development and human life depend on power system, this system also has the typical characteristics of big data. Power system is one of the most complex artificial systems, it has the characters of widely distributed geographical location, real-time balance of generating electricity, huge numbers of transmission energy, electricity transmission speed can reach, highly reliable communication scheduling, real-time operation and never stop, it expands in major failure immediately [2]. These characteristics determine the data that generated by power system in large numbers, rapid growth and type is rich, it is full compliance with all the features of big data, and it is typical big data.

© ICST Institute for Computer Sciences, Social Informatics and Telecommunications Engineering 2016
J. Wan et al. (Eds.): Industrial IoT 2016, LNICST 173, pp. 36–43, 2016.
DOI: 10.1007/978-3-319-44350-8_4

Under the situation of the smart grid deeply developing, it is more and more informationalized, intelligent and digital, and thus has brought more data sources, such as smart meters of terminals from millions of households and firms.

2 Processing Scheme of Big Data in Electrical Power System

In cloud computing mode, the power system of human-computer interaction and big data processing are both in the depth of fusion state. Therefore, based on cloud computing is the core of data processing, the key must be based on the integration thoughts, all of the network data cooperative organization, on the basis of integration, to maximize scatter power system of the comparative advantage of resources integration as a whole [3].

In cloud computing mode, the key to process huge amounts of data lies in the construction of power system, various kinds of scattered resources system and various other assistive technology. From a macro perspective, the whole big data processing can be divided into mixed processing and management. Mixed management is the key to realize the wireless power system, limited, resource sharing, scattered data management and the management synergy mechanism, etc. [2]; Mixing processing is the key to the fusion of large power system data processing operation model and relevant auxiliary technology [4].

At the same time, we need to focus on the redundant data processing. Redundant data processing scheme has the following several ways:

(1) only receive a set of data in a redundant data processing, abandoned another set;
(2) transfer redundant data to the higher level of data receiving equipment;
(3) filter redundant data as the unit of source equipment, if a set of equipment had any abnormal data we would take another set instead;
(4) receive the redundant data filtered according to the point as the unit.

3 Key Technologies of Big Data in Electrical Power System

3.1 Integration and Management Techniques of Big Data in Electrical Power System

The electric power enterprises' data integration management technology is to merge data from two or more application systems, to create a more enterprise application process. From the point of view of integration, it is the different sources and format, characteristics and nature of the data on the logical or organic concentration on the storage medium, providing the system of storing a series of subject-oriented, integrated, relatively stable, reflects the historical changes of the data collection, thus supplying comprehensive data sharing to the system [5]. Electrical power enterprises integrated management technology came into being is to solve the problem of the various systems of the internal data redundancy and information island between electric power enterprises.

Integration and management techniques of big data in electrical power system, including relational and non-relational database technology, data fusion and integration technology, data extraction technology, filtering technology and data cleaning, etc. An important feature of big data is diversity, this means that the data source is extremely widespread, data type is extremely complex, this complex data environment brings great challenge to big data processing, firstly it must carry on the extraction and integration of data source of data, and extract entities and relationships after association and aggregation, then unified structure is used to store the data, the data are required for data integration and extracting cleaning, guarantee the quality and reliability of data. Big data storage management is an important technology in no database technology, it adopts the distributed data storage method, and removes the relational characteristic of the relational database, data storage is simplified and more flexible and it has good scalability, which solving a lot of data storage problem. Representative no database technologies are Google's Big Table and Amazons chateau marmot [6].

3.2   Data Analysis Technology of Big Data in Electrical Power System

The fundamental driving force of big data technology is to convert the signal into data, to analysis the data of information, and to extract the information of knowledge, so that it can contribute to decisions and actions. With the big data analysis technology, we can find out the potential from the huge amounts of data in power system modal and law, and provide decision support for decision makers. McKinsey argues that the key technology that can be used for big data analysis is the result of statistics and computer science and other disciplines, including correlation analysis, machine learning, data mining, pattern recognition, neural network and time series forecasting model, genetic algorithm and so on.

Big data research study is different from the traditional logical reasoning, it is a huge amount of data to do a statistical search, classification, comparison, clustering analysis and induction, therefore inherited some characteristics of statistical science, such as statistics the data correlation or relevance of attention, the so-called "relevance" means two or more variables between the values of a certain regularity. The purpose of "Correlation analysis" is to find out hidden in the data set of networks, general with support, credibility, reflect the correlation parameters, such as interest in degrees [7].

Big data correlation analysis method, based on large amounts of sample, does not use the method of random such shortcuts, and adopts the method of analysis of all data; Big data of simple algorithm is more effective than small data of complex algorithm, the result is faster, more accurate and less susceptible to interference, we believe that based on correlation analysis of prediction is the core of big data.

Big data technology does not pay attention to the causality but focuses on the correlation analysis method, it has brought a big change in scientific research way of thinking, the late Turing award winner Jim gray data intensive scientific research, "the fourth paradigm" proposed by the big data research three former paradigm (theoretical science, computing science and experimental science) in isolated, alone. It can be a kind of research paradigm is because of the research methods of different from the traditional research method based on mathematical model [8].

Big data analysis technology in electrical power system, in essence, belongs to the traditional data mining technology in the new development of huge amounts of data mining, but the characteristics of huge amounts, high-speed growth, diversity determine it not only contains structured data, but also includes semi-structured and non institutional data, so many of the traditional data mining method of processing small data is no longer practical. Data mining and machine learning algorithms of big data in the environment of power system can be researched from three aspects:

(1) From the view of the management of big data and sampling and feature selection, changing the big data to small data;
(2) Do research about clustering, classification algorithm of big data, such as least squares support vector machine (SVM) based on conjugate degree (further squares support vector machine, the LS - SVM), random extensible Fuzzy C - Means (FCM), etc.;
(3) To carry out the big data parallel algorithm, through the parallelization of the traditional data mining methods, application to the knowledge of big data mining, such as machine learning and knowledge discovery based on graphs.

3.3   Data Processing Technology of Big Data in Electrical Power System

Data processing technology of big data in electrical power system includes distributed computing technology, memory computing technology, streaming technology, etc., these three technologies that can applicable objects and solve the main problem are shown in Fig. 1. Distributed computing technology is to solve large-scale data distributed storage and processing. Memory computing technology is effective to solve data reading and processing of online real-time computing. Stream processing technology is in order to get real-time, the speed and scale of uncontrolled data.

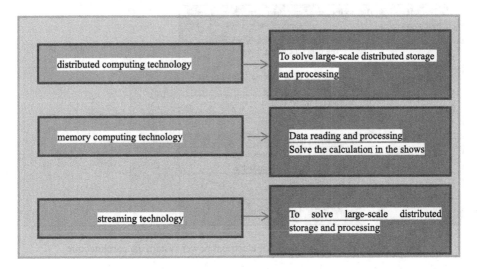

Fig. 1. Applicable object technology of data processing

In scattered data source of large-scale power system, we first use graphs programming model of distributed computing to block data that to be processed, then dividing into different Map missions, and the key-values stored in the local hard drive, finally we use the Reduce task to output the summary according to the key-value [9]. Memory computing technology puts all the data in the inner layer, it will overcome a lot of time consuming in the operations that disks read and write data, so the computing speed will increase greatly by several orders of magnitude. With the development of electric power industry, electric power system data quantity increasing, higher and higher to the requirement of real-time, applying data stream technology to power system can provide real-time basis for policy makers and meet the demand of real-time on-line analysis.

4   Electrical Power System Data Architecture Based on Cloud Computing

Cloud computing technology mainly has the following several aspects: data information network, service and customized and dynamic, etc., it has the characteristics that fully meet the power system's deployment and application of big data. According to the basic demand of large power system data applications, cloud computing technology can realize dynamic division or release of various resources, to realize the increase in the processing of resources, we can increase the resources available through proper matching to ensure that the resources are provided fast in the process of using elastic. If you no longer need to use the resources, it can realize the free release. Cloud computing technology configured dynamically, further implements the IT resources' continuous extension as needed (Fig. 2).

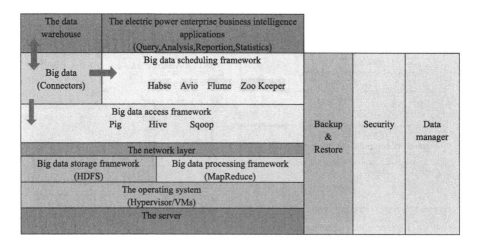

Fig. 2. The overall architecture of electrical power system data architecture

4.1   Network Architecture

By using cloud computing technology, to realize the change from traditional data physical model to the logical model, to realize the flexibility and agility of large power system in data processing through the construction of information network, to change the traditional IP network into SAN network, to change the big data processing network to carry on the further elaboration of layered, so that it can increase the elasticity of the network and flexible. Traditional physical network have been difficult to adapt to the requirement of cloud computing, virtualization, so by a variety of ways such as VXLAN, NVGER to change traditional physical network data flow type from two layers into three layers, to change the traditional physical layer into a logical layer in the network topology by increasing physical encapsulation at the same time, refining the classification management, so it can better meet the diverse needs from different types of needs from users [10].

Realizing the big data centralized control of the information network infrastructure. Through the automation network configuration of the scattered data, maximizing the efficiency of management and processing; Based on the data network virtualization processing to transit traditional physical network to the virtual abstract network, so that the application of the strategies of big data processing can be used to the workload. Comparing with the traditional processing way, this way has incomparable advantages: when the power load is deployed or migrated, the network configuration can be achieved automatically and adjusted to its settings.

4.2   Host Architecture

Host architecture plays a crucial role, it can be combined with bearing business characteristics to distinguish, to ensure that the system has reliable high-performance terminal, and effectively handle tightly coupled computing tasks through large-scale high-performance PC server or server cluster, such as large power system database, data mining, financial and marketing, etc.; Many general computing power system can provide low cost calculation and solution, especially in data processing of low hardware requirements of small and medium-sized application services business, this model not only can realize the processing of data, but also can effectively reduce the operation cost of the power industry through the adoption of high density and low cost intensive integration server cluster.

Host architectural pattern, on the other hand, has a fundamental change that from a traditional tower structure to a pool way. It can effectively achieve the host resources integration and optimization of the data processing system by adopting modern virtualization technology, modern virtualization technology and Rac One Node technology to realize database, such as composition and web level virtualization deployment and distributed among the Hadoop system, taking this model not only increases the speed of data processing, but also can effectively reduce the data processing system of all kinds of hardware failure of adverse effects on the business systems.

4.3 Storage Architecture

Storage system is the key to the whole system infrastructure and is the ultimate bearer of all the data in power systems. With the further development of various modern cloud computing, virtualization technology and big data technology, the traditional centralized storage has to be washed out gradually, and it is no longer the mainstream of data processing storage architecture. In order to deal with the access to huge amounts of data in electrical power system effectively, we must build the storage architecture with strong expansibility and scalability [11].

Cloud storage architectures based on cloud computing technology can effectively solve the problems that difficult to solve in the traditional architecture, it can store the data as cloud services. The key is to realize creation and distribution of huge amounts of data and gain the data through cloud service at the same time. The deployment of cloud storage technology mainly through the cluster or distributed file functions, the entire network of different types of storage devices in the system work together by related software, providing access to data storage and business, ensuring the safety of the whole system data meanwhile [12]. Through the distributed file system, such as object storage technology, providing scalable, extensible and strength data access function for all kinds of applications, meanwhile, as the adopted distributed technologies are based on standardized hardware technology, thus it can lower the cost of cloud storage and ops.

5 Conclusion

This paper expounds the big data processing scheme of big data platform in power system, and then expounds the key technology of four cores, namely the integrated management technology, analysis technology, processing technology, display technology. Comprehensive relational and non-relational database technology application should be considered and no database technologies should be emphasized on in big data integration management techniques; Carrying out the research of big data analysis techniques should be from big data's governance and sampling, large data's feature selection, big data smaller, data classification algorithm, parallel data mining, etc.; Big data processing technology should be taken into account according to the specific requirements of the application of distributed computing, memory, calculation, flow processing technology; Data display technology in electric power of big data can consider visualization technology, space information display technology and historical flow display technology. Finally, this paper illustrates the architecture of big data platform in power system, and gives overall executable frame of big data in power system, it has the reference value to the set up of data platform for electric power enterprises.

In the field of business, big data technologies have become more widely used and have created a huge commercial value, but its application in electric power system is just started, so combining technical advantage of big data and the application requirements of power system, exerting the value of big data in power system will bring new development opportunities for the smart grid construction. Power enterprises should firmly seize the opportunity, developing power big data technology from the aspects of data policy, talent training, key technology research and development, etc.

References

1. Shilong, L.: Grid integrated electrical equipment on-line monitoring platform based on the design of data acquisition system. Power Electr. **29**(2), 29–31 (2009)
2. Dewen, W.: Center infrastructure and key technologies based on the power of cloud computing data. Autom. Electr. Power Syst. **10**(11), 146–169 (2012)
3. Bingshuai, G., Jinlin, W., Xue, L.: A business related data collection method. J. xi'an Univ. Electron. Sci. Technol. **40**(2), 66–73 (2013)
4. McKinsey & Company. Big data: the next frontier for innovation, competition, and productivity **37**(1), 1–28 (2011). McKinsey Global Institute, New York
5. Wu, X., Zhu, X., Wu, G., et al.: Datamining with big data. IEEE Trans. Knowl. Data Eng. **26**(1), 97–107 (2014)
6. Kim, B.J.: A classifier for big data. In: Proceedings of the 6th International Conference on Convergence and Hybrid Information Technology. ACM, Daejeon (2012)
7. Kwon, T.H., Kwak, J.H., Kim, K.: A study on the establishment of policies for the activation of a big data industry and prioritization of policies: lessons from Korea. Technol. Forecast. Soc. Change **33**(20), 88–93 (2015)
8. Hansen, D.: Powering business analytics with big data and real-time using data integration. Database Trends Appl. **2**(1), 17–19 (2013)
9. Abdrabou, A., Gaouda, A.M.: Understanding power system behavior through mining archived operational data considerations for packet delivery reliability over polling-based wireless networks in smart grids. Comput. Electr. Eng. **18**(20), 118–119 (2015)
10. Avram, M.G.: Advantages and challenges of adopting cloud computing from an enterprise perspective. Procedia Technol. **27**(31), 199–201 (2014)
11. Divyakant, A., Philip, B., Elisa, B., et al.: Challenges and opportunities with big data. Proc. VLDB Endowment **5**(12), 2032–2033 (2012)
12. Shin, S.-J., Woo, J., Rachuri, S.: Predictive analytics model for power consumption in manufacturing. Procedia CIRP **29**(30), 191–194 (2014)

Research About Solutions to the Bottleneck of Big Data Processing in Power System

Ning Chen, Chuanyong Wang, Peng Han, Jian Zhang, Kun Wang, Ergang Dai, Wenwen Kang, Fengwen Yang, Baofeng Sun, and Guang Guo[✉]

State Grid Zaozhuang Power Supply Company, Zaozhuang 277100, China
43383047@qq.com

Abstract. The big data technology provides a new opportunity to the electric power system in various fields. In view of the shortage in traditional power system computing platform in computing, storage, information integration and analysis, this article puts forward a platform of power system based on cloud computing. Firstly, this paper outlines the development of power system can produce a large amount of data, and cloud computing technology is widely used in big data processing as a kind of new model, we can take chance of its application in power system. This paper discusses the relationship between cloud computing, big data and power system, and makes the conclusion that cloud computing technology can meet the demand of mass data storage and computing power system. Then we do research about the architecture, software technology of cloud platform system in electric power system. Finally, the feasibility of the application and development trend were discussed.

Keywords: Power system · Big data · Cloud computing

1 Introduction

With the development of long distance transmission system and the strength of inter-connected power system, large scale power systems have covered a number of countries continuously. The continuous expansion of power system and the complex structure is becoming increasingly difficult in safety evaluation, economic operation and system control. Electric power system will produce a large amount of data, the power system dispatching center should have powerful computation ability and information collection, integration and analysis capabilities in the future. The existing centralized power system computing platform is hard to meet the requirements, and it has become one of the main bottleneck of the realization of the smart grid.

Cloud computing is a new computing mode, and has gained rapid development in recent years, it includes several new computing technologies. So far, for cloud computing, there is no standard definition from authority. China's cloud Computing network cloud is defined as: cloud Computing is Distributed Computing, Parallel Computing and the development of Grid Computing, Grid Computing, or the commercial implementation of the concept of science. It is generally believed that cloud computing represents a large scale of distributed computing model based on the Internet.

© ICST Institute for Computer Sciences, Social Informatics and Telecommunications Engineering 2016
J. Wan et al. (Eds.): Industrial IoT 2016, LNICST 173, pp. 44–51, 2016.
DOI: 10.1007/978-3-319-44350-8_5

Cloud computing platform firstly using the Internet to a variety of wide-area and heterogeneous computing resources integration, in order to form an abstract and dynamically extension of virtual computing resources pool; And then through the Internet to users on demand provides computing power, storage capacity, software platform and applications, etc.

Through the establishment of power system, big data cloud computing platform can effectively integrate existing computing resources system, providing powerful computing and storage capacity to support a variety of analytical computing tasks. Cloud computing can support a variety of heterogeneous computing resources, compared with a centralized supercomputer, which can be highly scalable and can be easily upgraded in the existing computing power is insufficient. In addition, compared with the traditional computing model, cloud computing also has the ease of information integration and analysis, to facilitate the development of software systems, maintenance. In short, the establishment of the core system is calculated based on the power of cloud computing platform that can effectively address some of the key challenges in computing power system and information processing encountered in the future. This article will be defined for the cloud, features, technology, architecture, in power system research and application prospects and other issues were discussed in detail.

2 Big Data of Power System

Data generated by computing power system belongs to structured data, the main features are

(1) Multi-type isomers
     Different calculation software, different types of elements, different models of the same elements, and the results of different types of data structures are quite different;
(2) The homogeneity of calculating data online and offline
     Power System Simulation online data is usually calculated by putting the measurement information together with offline data, similar to the data structure, which has research data analysis on two kinds of data may be common;
(3) Diverse storage
     Online data is typically stored centrally, updated regularly, and most of the offline data are scattered in the staff's personal computer;
(4) Huge volume
     With the extensive use of smart grid scheduling technology support system (D5000), online data to calculate the rapid accumulation of volume will be achieved PB level [1]. While off-line calculation of the raw data is smaller, but huge volume of data analysis and calculation results are produced. For example, the entire network of 10,000 nodes grid bus get on short circuit current scanning, and calculate the whole network voltage of each bus fault under way will produce nine months results bus voltage 10, the data file reaches GB level [2].

As big data generated by power system has the main features above, we need a more efficient big data processing.

3   Power System Needs Big Data Cloud Computing Platform

Figure 1 briefly describes the relationship between power system, cloud computing, big data: Electric power system produces a large amount of data and needs to use cloud computing to save, processing and management the data; Cloud computing technology provides tools and methods to the development of big data technology, meanwhile, the latter provides application scenarios to the former, and both of them can share services for electric power system.

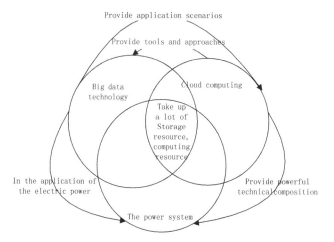

Fig. 1. Relationships among power system, big data and cloud computing technologies

Cloud computing can integrate intelligent power computing and storage resources, and improve power system and interactive processing capacity, becoming a strong technical composition of grid; Big Data technology and services based on business needs, rooted in the cloud computing; Cloud computing platform in power system can be considered to be big data applications in abstract, so relationship is between the three and they interact with each other [3].

The relationship between power systems, big data and cloud computing is a power system products of different stages of development from the more deep-seated, and it has heritage features.

Intelligent Power System is the result of information technology, computer technology, artificial intelligence technology in the traditional power system applications precipitation, meeting the power system information, intelligence, clean high-level operational and management needs, not only on the traditional grid inheritance, but also to carry forward the traditional power grid, so its development must be synchronized with the new technology. Forefront of cloud computing and big data technologies from most computer and information technology is their stage of development and technical application level that has a landmark new technology [4].

Cloud computing technology in the distributed storage technology and parallel computing technology, satisfying the mass data storage and computing power system requirements, so power cloud is proposed soon after the cloud computing technology, the concept of cloud computing technology application in power system is also shows a tendency of flowers gradually, of which promoting the development of the power system. Big data technology is not only the continuation of traditional data analysis and mining technology, but also an inevitable product of the knowledge mining and business application demand, so most of the applications of big data technology based on distributed storage and processing technology with the key technology of cloud computing. In a sense, the development of the power of big data technology can be regarded as cloud computing technology in power system and the implementation process of high-level business requirements.

4   Cloud Platform for Processing Big Data in Power System

The power systems cloud computing platform is achieved by architecture and technologies, and is a complex entity composed by a variety of devices and users that connected to each other through Internet component (see Fig. 2). Generally speaking, cloud computing platform can be divided into two main parts, namely the control center and a variety of computing resources integrated by cloud computing platform. The main function of the control center in cloud computing is based on the user's request, dividing user's computing tasks into several sub-tasks, and then dynamically assigning sub-tasks by the Internet to the computing devices that integrated by cloud computing platform [5]. After each sub-task is completed, its results will be re-aggregated to the control center via the Internet, and finally back to the user.

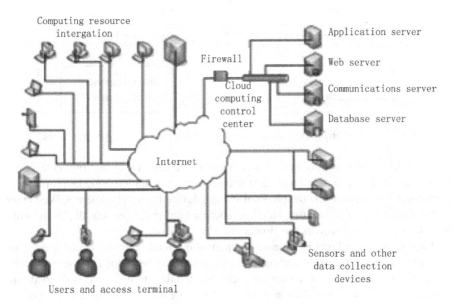

Fig. 2. Structure of cloud computing

Cloud computing platforms and data acquisition are connected via Internet network and a large number of sensors and other data collection devices. For power systems, future data collection network not only includes traditional SCADA system sensor, but also PMU and placement of smart meters in the end-user at home, or even a variety of smart appliances embedded systems. These devices can provide a full range of system information. In addition, the power system can also connect cloud computing platform and other data sources, such as: regional meteorological database connected to each other, to get the temperature, humidity, wind speed, sunshine and other data. The amount of data generated by such a large-scale network collected would be amazing, it can be stored and analyzed only by virtue of cloud computing platform's computing power. Taking into account a lot of power system analysis tasks on time-critical, you should consider creating a dedicated high-performance network to connect to cloud computing platforms and data collection network, to improve the reliability of data transmission [6]. Power Systems cloud computing system architecture platform as shown in Fig. 3.

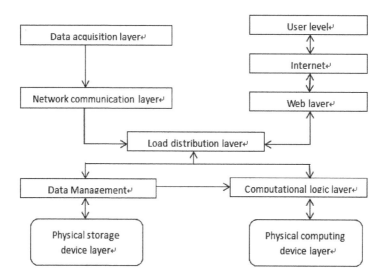

Fig. 3. System architecture of the cloud computing platform for power systems

From the system architecture point of view, cloud computing platform consists of Web layer, load distribution layer, data management, calculation logic layer, physical computing device layer and the physical storage device layer [7]. Wherein, Web layer is responsible for implementing cloud computing platform, Web site that is the only interface to the user to access the cloud computing platform.

Load distribution layer is the core component of cloud computing platforms. This layer has four main functions: computing tasks of user will be divided into several parts, and decided to enable what device for each task; the data to be stored is divided into several parts, and determined the appropriate storage device; computing and integrating results the logic layer returned, and then feedback it to the user; according to the read data request, the data management layer reads data and then outputs it after integration.

The four layers above consist of the software part of the cloud computing platform. Physical computing device layer and the physical storage device layer represents all the physical devices that make up the hardware part of the cloud computing platform [8].

The following discusses several important software technology can be used to achieve cloud computing power system.

(1) An important feature of the service-oriented architecture of cloud computing is online expansion and upgrades, which puts forward higher requirements of flexibility for the software [9]. Service-oriented architecture is a rapidly growing software design method in recent years. In the traditional design methods and software to function or class as a basic function module to the application programming interface (API) as a means of communication between different programs, SOA is to serve as the basic functional modules. Compared with the function, service representative higher-level application requirements (for example: to print the data from the database and to read the report of the entire process can be abstracted into a service, but to read the database is only a function). According to user needs, each one of the main functions is packaged in the form of services by SOA, and each service is independent, only by Extensible Markup Language to realize communication [10]. When any kind of function needs to be updated, you just need to replace the corresponding services. In addition, SOA-based architecture can freely combine a number of services to the rapid formation of a new system. For example: it can package flow calculation as a service, so there is no need to do the task flow calculation, what should be done is to tide computing services and other related services to by online combination. Visibly, SOA applications can greatly improve the flexibility and software systems to upgrade the speed of development. Of course, raise the level of abstraction of software generally at the expense of the efficiency of communication. Summarize the history of software development methods, it develops from process-oriented to object-oriented, and then to service-oriented. Software development has trend to improve the level of abstraction overall, and this is compatible with the software itself increasing complexity. With the continuous development of power system, the function of power information system growing, and structure of itself is increasingly complexity, which will bring difficulties to developers who use traditional development methods. Therefore, raising the level of abstraction of software is an inevitable trend. For communication efficiency, you can define a service level of abstraction appropriate to strike a balance.

(2) Dynamic load distribution

Load distribution algorithm is the core of cloud computing. Dynamic load distribution has proven to be a class of more effective allocation algorithm for computing tasks. The basic principle is to allocate computing tasks based on the calculation speed of each device dynamically; computing device of greater speed is assigned the bigger task, of which to ensure all computing devices simultaneously return results. For large-scale cloud computing platform, you can also consider using the job replication methods to improve reliability, it is about to copy a number of copies of each sub-task, and distribute them to multiple computing devices to synchronously processing, thus avoid slow the overall efficiency down by the occurrence of a device failure that results in redistributing sub-tasks.

In addition to the discussion above, there are still a variety of commercial or open source software technologies can be used to cloud computing, such as GoogleMapReduce, Google File System (GFS), Microsoft's Dryad/DryadLINQ, and other opensource distributed computing framework [12]. Considering the power system is an important national infrastructure, we think open source software technology should be the foundation to realize cloud computing platform in power system.

5   Prospects

With the development of technology, electricity big data platform will be similar to EMS, ERP and other systems, and becoming an "infrastructure" of planning, design, operation and management for the power system in the near future. Perspective in all areas of power system will change from current information and results to trends against data traceability to the source of the problem. At the same time, it will comply with the variation of the data, explore new trends, and find hidden in one of the next scene.

Cloud platform for big data processing power system, as a guarantee for validity of the data processing, all the research we do need to have a solid theoretical foundation. Thus, the research should from the discovery of phenomenon to the analysis of phenomenon to discovery of reasons behind it, and rise to the level of theory as far as possible, which brings a new challenge for applying cloud computing technology to the applications.

In summary, the combination of power Big Data technologies and Cloud Computing will be able to provide a new means of system analysis, perspective, and even methods.

References

1. Kwon, O., Lee, N., Shin, B.: Data quality management, data usage experience and acquisition intention of big data analytics. Int. J. Inf. Manag. **18**(1), 156–157 (2014)
2. Yaqi, S., Guoliang, Z., Yongli, Z.: Present status and challenges of big data processing in smart grid. Power Syst. Technol. **37**(4), 927–935 (2013)
3. Xiaofeng, M., Xiang, C.: Big data management: concepts, techniques and challenges. J. Comput. Res. Dev. **50**(1), 146–169 (2013)
4. Divyakant, A., Philip, B., Elisa, B., et al.: Challenges and opportunities with big data. Proc. VLDB Endowment **5**(12), 2032–2033 (2012)
5. Meng, L., Xiaodong, C., Wen, Z., et al.: Design of cloud computing architecture for distributed load control. Power Syst. Technol. **36**(8), 140–144 (2012)
6. Xiaoyang, T., Shengyong, Y.: A survey on application of data mining in transient stability assessment of power system. Power Syst. Technol. **33**(20), 88–93 (2009)
7. Chen, C.L.P., Zhang, C.-Y.: Data-intensive applications, challenges, techniques and technologies: a survey on big data. Inf. Sci. **14**(9), 118–119 (2014)
8. Ni Zhang, Yu., Yan, S.X., Wencong, S.: A distributed data storage and processing framework for next-generation residential distribution systems. Electr. Power Syst. Res. **19**(26), 77–80 (2014)

9. Prasad, A., Green, P., Heales, J.: On governance structures for the cloud computing services and assessing their effectiveness. Int. J. Account. Inf. Syst. **15**(4), 335–356 (2014)
10. Zhao, X., Zhou, C., Zhao, L.: Condition evaluation model of fluid power system in gradual failure based on data envelopment analysis. Comput. Fluids **27**(31), 199–201 (2014)
11. Pereira, L.E.S., da Costa, V.M.: Interval analysis applied to the maximum loading point of electric power systems considering load data uncertainties. Int. J. Electr. Power Energ. Syst. **9**(14), 97–99 (2014)
12. Shin, S.-J., Woo, J., Rachuri, S.: Predictive analytics model for power consumption in manufacturing. Procedia CIRP **29**(30), 191–194 (2014)

Research of Mobile Inspection Substation Platform Data Analysis Method and System

Peng Li[✉], Ruibin Gao, Lu Qu, Wenjing Wu, Zhiqiang Hu, and Guang Guo

State Grid Jinzhong Power Supply Company, Jinzhong, China
5259355@qq.com

Abstract. With the extensive construction of mobile inspection substation plat-form, various kinds of substation information is going to be digitized into data-bases. The scale of data is increasing day by day, which is large enough to provide a data base for data mining technology. Association rules has been applied successfully in various fields which makes it as one of the most active branches in the field of data mining research, development and application. Association rules used in smart substation data analysis to find some rules that people cannot find easily and it can find laws of equipment aging and failures also. This will benefit substation management, as well as equipment maintenance. In this article, we present a data analysis system based on mobile inspection substation platform. And the improved Apriori algorithm is used in substation data analysis to dig out some of the basic laws which provide effective information for substation management.

Keywords: Mobile inspection substation platform · Data analysis · Data mining · Association rules · Improved Apriori algorithm

1 Introduction

Substations play a key role in the sphere of power transmission and information collection. They generate large amount of data which ensure informationization of Mobile inspection substation platform. Substation applications also rely on data integration system. That's why data analysis is essential.

Development of computer technology and increase in demand for electric power system promote standardized management of substation. With the research of smart grid developing throughout the world, standardization of information has become a key technology of the future development of the power system. Construction of standardized data centers which has information on achieving data integrity substation, consistency, accuracy and standardization is very meaningful. It will also promote research on substation data analysis. Gradually building and development of mobile inspection substation platform will produce millions of data in a single day, which would provide data foundation for data analysis applications. Association Rules (AR) becomes one of the most active branches of data mining research, development and application [1]. There are a lot of researchers have been involved in the association rules on the issue of data mining research.

© ICST Institute for Computer Sciences, Social Informatics and Telecommunications Engineering 2016
J. Wan et al. (Eds.): Industrial IoT 2016, LNICST 173, pp. 52–58, 2016.
DOI: 10.1007/978-3-319-44350-8_6

In this paper, a mobile inspection substation platform data analysis system contents data acquisition, data storage, data analysis and application of analytical results. The proposed implementation method of substation data analysis is based on an improved Apriori data analysis algorithm. It combines a variety of professional data and discovers the hidden information of substation data and brings convenience to the operation and management of substation.

2   Mobile Inspection Substation Platform Data Analysis System

Mobile field work platforms have changed the traditional working methods of artificial paper registration procedure by using the mobile terminals. This can improve the efficiency and quality inspection, electronic of substation equipment inspection. It can also minimize mistakes, ensure inspection officers work effectively.

Firstly, the platform collects data through the mobile terminals and storage it into database. Secondly, the system finds the hidden laws by mining and analyzing the data in the database and sends the results to management Platform. Thirdly, the substation management personnel screens the results and selects effective ones. Finally, the effective results are used in guiding new inspections.

The mobile inspection substation platform data analysis system is divided into four parts. They are mobile terminals, management platform, server and database [2] (Fig. 1).

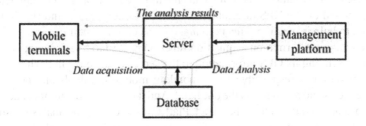

Fig. 1. The system block diagram

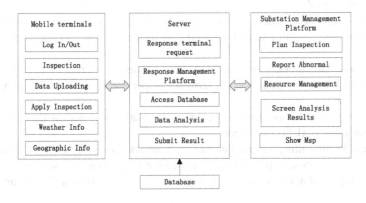

Fig. 2. The system function block diagram

The mobile terminals are handheld devices for inspection officers, who are responsible for the inspection of substation equipment. They can ill inspection report or put forward inspection application [3]. These terminals are data capture terminals. The mobile terminals interact with the server in the form of 3G network generally. The mobile terminal subsystem includes log in and out function, inspection function, Data uploading function, apply inspection function, weather Information function and geographic information function. The most important ones are the inspection and data upload functions which include filling weather information, preparing tools, confirming the danger points, filling inspection results, taking pictures of evidence and uploading inspection results (Fig. 2).

The log in and out function provides an authentication mechanism to ensure that just the users who are accepted by system are allowed to log in. It also provides a convenient way to send worksheets to the correct staff.

The inspection and data uploading function is the most important function of the mobile terminal. Inspection staffs complete the inspection task through the mobile terminal. They fill weather information, prepare tools, confirm the danger point, fill in the inspection results and take photo of evidences. Then they send the inspection results to the server.

The inspection task application function makes the inspection staff be able to ask for a new inspection task by the actual situation. But they must offer the inspection date, type, substation and reasons.

The weather information function means the mobile terminal system can determine its location based on the GPS information to get the locate weather information which can be used to help inspection officers to make plans of inspection tasks.

The geographic information acquisition function is used to track terminal trajectory Real-time by sending its GPS information to the server every once in a while.

Platform server is responsible for responding mobile terminals and management platform requests and responds on the one hand. And it is responsible to store and retrieve database to analysis data on the other hand. Platform server gets the analysis results and submits the results to management platform [4]. Its main features include responding terminal request, responding management platform, accessing database, analyzing data and submitting results.

The responding to mobile terminals' requests function is designed for mobile terminals applying inspection tasks and uploading inspection results and geographic information.

The responding to management platforms' requests function is used to accept requesting of issuing worksheets, get terminal inspection result data and geographic information data.

The accessing database function refers to the platform to access the database. It includes database CRUD operations.

The results submitting function is used to feedback data analysis results. Database is responsible for various types of data. Management Platform is used to develop inspection plans, report abnormalities, manage resource, screen Analysis Results and show Maps.

The inspection plan management function contains making, reviewing, modifying, deleting and approving plans. And the data analysis results can be used in plan making.

3   Data Analysis Method of Mobile Inspection Substation Platform

3.1   Data Mining Algorithm and Association Rules

Association rules are mainly used in finding associations of transaction database items or attributes. It focuses on identifying the set of attributes frequent and frequent itemset from data. Then it uses these to create the association rules. Association rules are not based on the intrinsic properties of the data itself. It is based on characters of the simultaneous occurrence of the data items. Association rules are succinct and understandable.

The basic model of association rules is as follows. The problem of association rule mining is defined as: Let $I = \{i1, i2,..., in\}$ be a set of n binary attributes called items. Let D be a set of transactions called the database. Each transaction in D has a unique transaction ID and contains a subset of the items in I. A rule is defined as an implication of the form $X \Rightarrow Y$ where $X \subseteq I$, $Y \subseteq I$ and $X \cap Y = \emptyset$ [2]. The sets of items (for short itemsets) X and Y are called antecedent and consequent of the rule respectively.

To select interesting rules from the set of all possible rules, constraints on various measures of significance and interest can be used. The best-known constraints are minimum thresholds on support and confidence. The support $supp(X)$ of an itemset X is defined as the proportion of transactions in the data set which contain the itemset. The confidence of a rule is defined $conf(X \Rightarrow Y) = supp(X \cup Y)/ supp(X)$. The association rules algorithm follows the general process.

Among so many association rules algorithms, Apriori algorithm, which was proposed in 1993 by R. Agrawal is one of the most influential ones. Apriori is designed to operate on databases containing transactions. Apriori Association rules consist of the following steps [5] (Fig. 3).

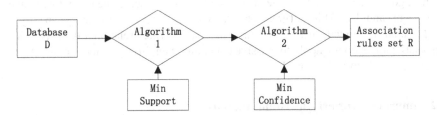

Fig. 3. The general process of association rules algorithm

Step 1, According to user-set minimum support frequent item sets, priori iterates to identify all frequent item sets. Step 2, It gets strong association rules by the minimum degree of confidence. The process of Apriori algorithm is as Fig. 4.

Apriori algorithm has its fatal deficiencies. First, when each element needs to decide whether to join frequent item sets by verifying, it cause the algorithm to scan the database multiple times, which will cause excessive I/O operations. This reduces the efficiency

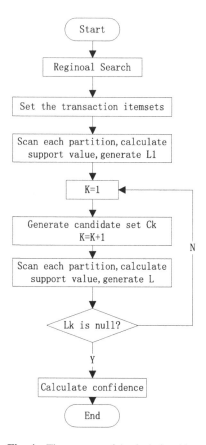

Fig. 4. The process of Apriori algorithm

of the algorithm. Second, when the project set has a larger scale, the project will have a huge set of candidate. This will greatly consume system's CPU and memory capacity [6].

It cannot be directly used for mining association rules relational databases. This association rules do not apply to massive data mining environment too. So, the Apriori algorithm needs to be improved.

3.2 Improved Apriori Algorithm Analysis

Currently, there are a lot of association rule algorithms. The vast majority are based on Apriori algorithm as a basis for optimization improvements. They reduce I/O operations by reducing the number of scanning database to improve the efficiency of the algorithm.

By combines Sampling algorithm proposed by H. Toivonen and DIC algorithm proposed by S. Brin et al., we subdivide substation database, and then sample it to improve the efficiency of obtaining frequent item sets [7].

Substation operation information data distinguish in accordance with the substation node. Any association rules are only valid within the same substation. This is consistent with DIC algorithm which demands for data partitioning [8]. The process of Improved Apriori algorithm is as follows (Fig. 5).

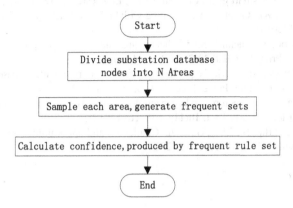

Fig. 5. The process of improved Apriori algorithm

The improved Apriori association rules proposed by this paper require a combination of experience of substation staffs. This mining that the accuracy requirements of this algorithm can be slightly reduced [10].

4   Conclusion

In this paper, the design of data analysis system based on mobile inspection substation platform is studied. And the association rules data mining technology is introduced into substation data analysis which based on the data characteristics of mobile inspection substation platform. We apply the improved Apriori association rules algorithm to substation data analysis. A smart substation data analysis system has been established. With the rapid development of smart grid, in particular, the promotion and construction of substation integrated digital technology, this system will providing a wide application platform for growing amounts of substation data.

References

1. Zhang, P., Gao, X.: Mobile inspection substation platform. Electr. Technol. **2010**(08), 31–40 (2010)
2. Berzal, F., Cubero, J.-C.: An efficient method for association rule mining in relational databases. Data Knowl. Eng. **37**, 47–64 (2001)
3. Chen, F., Deng, P.: A survey of device collaboration technology and system software. Acta Electronica Sinica **39**(1), 440–447 (2011)
4. McKinsey & Company. Big data: The next frontier for innovation, competition, and productivity, vol. 37(1), pp. 1–28. McKinsey Global Institute, New York (2011)

5. Abdrabou, A., Gaouda, A.M.: Understanding power system behavior through mining archived operational data considerations for packet delivery reliability over polling-based wireless networks in smart grids. Comput. Electr. Eng. **18**(20), 118–119 (2015)
6. Wu, X., Zhu, X., Wu, G., et al.: Data mining with big data. IEEE Trans. Knowl. Data Eng. **26**(1), 97–107 (2014)
7. Shin, S.-J., Woo, J., Rachuri, S.: Predictive analytics model for power consumption in manufacturing. Procedia CIRP **29**(30), 191–194 (2014)
8. Lin, S.: Grid integrated electrical equipment on-line monitoring platform based on the design of data acquisition system. Power Electr. **29**(2), 29–31 (2009)
9. Wang, D.: Center infrastructure and key technologies based on the power of cloud computing data. Autom. Electr. Power Syst. **10**(11), 146–169 (2012)
10. Divyakant, A., Philip, B., Elisa, B., et al.: Challenges and opportunities with big data. Proc. VLDB Endowment **5**(12), 2032–2033 (2012)

Data Recovery and Alerting Schemes for Faulty Sensors in IWSNs

Huiru Cao, Junying Yuan[(✉)], Yeqian Li, and Wei Yuan

Department of Electronic Communication and Software Engineering,
Nanfang College of Sun Yat-sen University, Guangzhou 501970, China
caohuiru0624@163.com, cihisa@outlook.com,
{liyq,yuanw}@mail.nfu.edu.cn

Abstract. In monitoring and alerting industrial system, industrial wireless sensor networks play an important role. However, we usually have to face one critical issue that is to recover the data and emergency treatment schedule for the faulty sensors. In this paper, we target on monitoring industrial environments and deal with the problems caused by the failure or faulty sensors nodes. Firstly, based on industrial private cloud, an architecture of industrial environment monitoring system is proposed. Furthermore, a hierarchical support vector machines is adopted for faulty nodes' data recovery. Unlike most previous works, we intend to address the problem from global and local data perspectives. Using the first layer Support Vector Machines is adopted to judge the types of missing data based on the monitoring system. In second layer of SVM is responsible for finishing the recovery local data in the light of the history records. Performance of the proposed SVM data recovery strategies are evaluated in terms of networks self-healing competence, and energy consumption. We also implement our schemes in a real-life monitoring and alerting network system to demonstrate the feasibility and validate the network detection capability of emergency events.

Keywords: Data recovery · Hierarchical support vector machines · IWSNS · Alerting system

1 Introduction

In the past decade, with the increasingly developments of information, computer and communication technologies, and wireless sensors networks (WSNs) have significantly encouraged advances. Because in a WSN there are many advantages such as no wires, flexibility, mobility, low cost etc. Furthermore, in different applicable fields for WSNs, many challenges have to face for academic and engineering. So WSNs become a hot issue in research domain. Meanwhile, we usually encounter that a WSN is widely used for smart home, medical health, military fence and agriculture [1–4]. In these WSNs application, environmental monitoring and alerting system occupy a great share.

In industrial fields, usually WSNs are adopted to monitor environment performances such as temperature, humidity, light intensity vibrations, heats, noise and other parameters [5–9]. Moreover, IWSN working environments are not friendly and full of

© ICST Institute for Computer Sciences, Social Informatics and Telecommunications Engineering 2016
J. Wan et al. (Eds.): Industrial IoT 2016, LNICST 173, pp. 59–69, 2016.
DOI: 10.1007/978-3-319-44350-8_7

water, metal, high humidity and temperature. It is easy to cause node failure and monitoring data missing. If the vital monitoring sensors nodes are lose their functions such as communications, sensing data, the imponderable disaster will happen, and there are amount of economic loss [10, 11].

As shown in Fig. 1, in a monitoring IWSNs system, wireless sensor nodes are deployed in an industrial domain. It is common that sensor nodes are placed with maximums coverage as Fig. 1(a). In the industrial environment, as above discussing wireless sensor nodes have to face many challenging. So nodes failures are very common like Fig. 1(b). Once the faulty nodes exist in WSNs, it maybe forms some sensing void holes for example in Fig. 1(c). If the emergent events happen in area of the sensing holes, it must be unreliable for this kind system. How to recover the missing data and emergency treatment of faulty nodes in the sensing void holes become an important way to deal these challenge with the lowest cost and real time. These are motivation and goal problem of this paper.

In this work, we do not intent to study the improvement of processing of node design for overcoming the hostile environment or longing network working life, nor nodes deployment with amount of sensor to achieve certain degree of sensing coverage. Rather, based on the historical record of global and local sensing data, we are from the data prediction perspectives to deal with the urgent event, and avoiding this disaster caused by node failure. Our ultimate goal is to realize industrial WSNs for monitoring and alerting system so that detection application of various emergency events with few nodes faulty can be practically implanted.

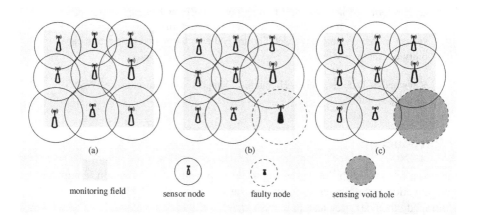

Fig. 1. Sensing void holes in IWSN

The remainder of the paper is organized as follows. In Sect. 2, related works are introduced for solving the faulty sensor nodes. Also, several prior research efforts and summarize our works contributions. Then, Sect. 3 give architecture of monitoring system based on Industrial private cloud and Industrial WSNs. In Sects. 4 and 5, the hierarchical SVM data recovery and emergency treatment strategies are proposed.

Meanwhile, the performance and comparison are presented, while a real-life industrial monitoring network are implemented for demonstrating the proposed schemes. At last, we draw our concluding remarks in Sect. 6.

2 Related Work

For finishing the goal (environment monitoring), most of earlier study works in WSNs applications and researches, focus on the low real-time environments such as greenhouses, forests, refineries and so on. It is known that these fields are not easy to access and nodes are amounted limited energy [12, 13]. These kinds of WSNs have to meet remote communication, large-scale monitoring range, especially for saving energy of the whole networks. In many related works, the energy conservation schemes are proposed. Among them, the most classical strategies that ordinary nodes switch between sleep and waking, extending sleeps time, or turn off the wireless radio module. It is obvious that these schemes are not suitable for industrial applications because of higher real-time and being sensitivity to emergencies.

With the deepening of WSN research, it is possible to deploy node with some effective and special schemes. So in some literatures [14, 15], various deployment strategies have been presented to enhance the sensing void hole. One important and common way is that deploying redundant nodes in monitoring fields. It get an effective means to deal with this challenge. However, such redundant deployment means higher cost and it is hard to judge the position for placing the redundant nodes. Furthermore, it is vulnerable to sensing void holes, particularly in industrial domain.

Consequently, a plenty of research efforts begin to transfer to the mobility WSNs, by mobile sensors nodes to large sensing range and fill the sensing nodes, given any number of randomly placed sensors nodes. However, in these WSNs, for finishing effective moving the networks have to transfer related data or information to all moving nodes. Moreover, a lot of energy must use to finish mechanical moving.

We observe that previous works explore these strategies only partially, leaving issues such as energy conservation and high latency, data recovery, sensing void holes, dealing the emergent event. However, in practice, data recovery of sensor failures should be resolved as an effective solution to achieve an operative WSN with high self-healing for void holes. In light of this, we investigate the sensor data, and based on Global SVM divided the data into different type for meaning the different level of industrial condition. Based the value of different level, IWSNs could drive different services. Then, for getting missing data of sensor nodes failure, local SVM is adopted to forecast the single nodes missing data. We summarize our contributions as follows. Firstly, we develop a double SVM for dealing sensing void holes. Secondly, our system could quickly respond to emergent event. At last, a real-life WSN adopting our strategies are implanted.

3   Industrial WSNs Architecture for Monitoring and Alerting System

It is known that WSNs having many advantages are applied in industrial domain. Among of these IWSN, most of applications is targeted to monitoring industrial environment and machine conditions. Once the sensing environment parameters are beyond the normal ranges, the alerting systems are driven, and some appropriate measures will be taken, such turning off power, evacuating the workmen etc.

3.1   Architecture of Industrial WSNs for Monitoring

As above discussing, a complete industrial monitoring system has several functions: monitoring, processing data, judging urgent event and give corresponding warning. As Fig. 2 shown, in our industrial WSNs monitoring and alerting system architecture, the framework contains four parts: sensor nodes, backbone networks, data servers and services.

In an industrial monitoring and alerting system are based on WSNs. The nodes are deployed in different places as certain schemes. Sensor nodes collect the data, periodically. Then the nodes send the node to sinker or base stations, according to certain communication protocols such as MAC. In the paper, TDMA is adopted to transmit the sensing data. Thereafter, sinking data are unloading to servers or industrial private clouds. At last, some relevant services were supported from the server to mobile users, workmen and management system.

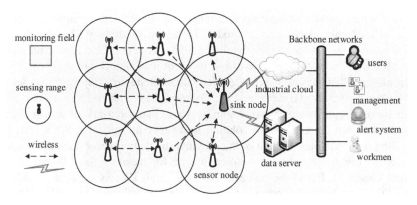

Fig. 2. Architecture of monitoring and alerting system in IWSN

3.2   IWSNs Monitoring Working Algorithm with Dealing with Emergent Events

It is known that in an industrial environment emergency mechanism for emergent event such fires, accidents are very important. A good IWSNs monitoring system should have an effective way for dealing with the emergent events. Usually, the traditional monitoring system would transmit the alerting information after the emergent event taking place. Obviously, it is not suitable for the modern industrial environment system. So in this section, we give an algorithm for deal with the emergent events based on the prediction with the historical data records about the monitoring system.

Table 1. Summary of notations used in IWSNs monitoring working algorithm

Notation	Description
k	Loops of sensor nodes worked
T_s	Time of sensor nodes start working
T_{cyc}	Time of sensor nodes cycle working
Tc	Time of current
S_i	The ith sensor nodes
$D_{i,k}$	The ith node sensing data in the kth loops
Max	Max loops
e_i	The ith emergent event

Table 1 gives the description of notation of our IWSN working steps. Algorithm 1 provides the pseudocode for IWSNs monitoring operations. Note that in the end of the k-th loop, sensor node Si sensing data Di, k is judged whether it is beyond the emergent set $\{e1, e2, e3, ..., en\}$. Then system performs the alerting or normal operations, based on the judged results. For giving a clear description of IWSNs monitoring system, we give an algorithm about the working processing. The system working principle as following:

Algorithm 1. IWSNs monitoring working algorithm with dealing with emergent events.

1:	set $k = 0$;
2:	set $T_s = T_{cyc}$; //initial sensing time
3:	**while** $(k < Max)$ && $(Tc = Ts)$ **do**
4:	**for** each sensor $s_i \in \{s_1, s_2, s_3, ..., s_n\}$ **do**
5:	sensing data;
6:	transmit data to servers or cloud by WSNs.
7:	**End for**
8:	Server saving and judging these sensing data $D_{i,k}$
9:	**if** $(D_{i,k} \in ! \{e1, e2, e3, \cdots, en\})$ **then**
10:	perform normal strategies
11:	**else**
12:	perform alerting strategies
13:	**end if**
14:	set $k = k+1$;
15:	**end while**

---

4 Hierarchical SVM for Node Faulty Data Classification and Recovery

Wireless sensors nodes operate in industrial fields have to face different formidable challenges, such as signal interferences, obstacles, dusts, high humidity and temperature. So wireless sensors are inherently unreliable. The sensor depletions or unexpected failures will cause missing data, disaster and sensing void holes. To prevent accident taking place and recover the missing data, one alternative namely hierarchical SVM strategy is to perform.

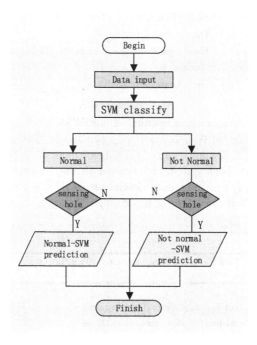

Fig. 3. Hierarchical SVM for dealing with data classification and data recovery

In this section, we use hierarchical SVM for dealing with data classification and data recovery. As in Fig. 3, the first one-class SVMs classifier will be used to classify the environment of industrial working fields. Usually we divide the data in safe and unsafe, it is mean that no emergent event and emergent event happen in the monitoring domain. If the domain has emergent event, the handling mechanisms are triggered. Secondly, as soon as there are faulty nodes (sensing holes) in the monitoring area, the mechanisms of data prediction are called for. The data prediction is based on the SVM and historical data records. SVM prediction uses the SVM classifier to judge the previous data of faulty nodes to predict.

4.1   Global SVM Data Classification

In the present context, one-class SVMs classifier will be used to detect monitoring data in an industrial sensor network. SVM classifier could divide these data into different level of the monitoring industrial field. One-class approach based on the fact that usually monitoring system draws more attention to whether the monitoring environment is in safety. Hence the output set have two values, not-normal and normal. As Fig. 4 shown that for simply to compute, we use +1 to label normal and −1 to label not-normal. The SVM working mechanism is to search the optimal plane for divide the data in different classifications.

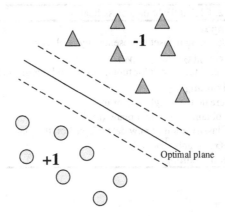

Fig. 4. Linear separation of the industrial monitoring data into two classes

Related SVM problem is formulated as follows: Given that the training set is

$$T_{\text{set}} = \{(x_1, y_1), (x_2, y_2), \ldots, (x_N, y_N)\} \in (X \times Y)^N$$

$$\text{where } x_i \in R^{dL}, \ i = 1, 2, \ldots, N$$

And X denotes the input space, Y denotes the output space. d_L is the dimension of input space. N is the size of training set. In light of this training set a decision function is constructed.

$$y(x) = \text{sgn}\left[\omega^T \varphi(x) + b\right] y(x) = \text{sgn}\left[\omega^T \varphi(x) + b\right]$$

Where ω is the weight vector and b is the bias. This may construct the one-class SVM primal problem.

$$\min_{\omega, b, \varepsilon} f(\omega, b, \varepsilon) = \frac{1}{2} \|\omega\| + C \sum_{i=1}^{n} \varepsilon_i + b$$

$$s.t. \begin{cases} \omega^T \varphi(x_i) + b \geq -\varepsilon_i \\ \varepsilon \geq 0 \end{cases}$$

Where C is some constant and ε is a vector of slack variables. It is known that the first term is a regularization term for preventing overfitting. The second is an empirical risk estimation function, and the final term is included to bias the result to detecting the industrial working environment. For giving a clear description of global-SVM data classification, we give an algorithm about the working processing. The system working principle as following:

Algorithm 2. Global SVM data classification.
1: **Input:**
2: //training dataset represented by N blocks:
3: $(x_1,y_1),(x_2,y_2),\ldots,(x_N,y_N)$
4: // constant $C > 0$ for tuning errors and margin size
5: **Training:**
6: -create the weight vector ω
7: - obtain the optimal plane (ω, b)
8: Classification of new data x based on the plane is:
9: $y(x) = \text{sgn}[\omega^T \varphi(x)+b]$
10 **end Training**

4.2 Local SVM Data Recovery

For deal with the sensing data holes in industrial WSN monitoring system, we propose a SVM algorithm to predict the missing data. The whole working steps of local SVM data recovery is divided into two stages. Table 2 give the notation used in algorithm 3.

Table 2. Summary of notations used in Algorithm 3

Notation	Description
i	ID of faulty nodes
D_{now}	Current sensing data of node
D_j	Previous historical data of faulty nodes
Nor_i	The ith neighbor nodes of faulty nodes
D_{miss}	Missing data
S_{sink}	Sink sensor nodes
$D_{i,k}$	The ith node sensing data in the kth loops
y	The result of algorithm 2
m_{Max}	Max allowing cycle of the faulty nodes stop working
m	After the m-th sensing cycle of the faulty nodes stop working

First, the WSNs looking for faulty sensing nodes, and transmit the nodes is to server or industrial private clouds. Second, the server predicts the monitoring area whether to normal or not-normal based on the Algorithm 2 with previous historical data. Furthermore, based on the result of above, the server begins to predict the data using SVM. At last, the prediction data are transmitted to nodes or sink and neighbor nodes.

Algorithm 3. Local SVM data recovery when faulty sensing data stop working.	
1:	set $m = 0$;
2:	Sensor node transmitting data to sink S_{sink}
3:	Sink nodes inform the server or clouds
4:	**if** ($D_{now} = void$) **do**
5:	input: $D_1, D_2 D_3, ..., D_j$
6:	call for the Algorithm 2.
7:	output y;
8:	**while** ($m < m_{Max}$)
9:	**if** ($y = +1$)
10:	D_{miss} = SVM for normal prediction
11:	**else**
12:	D_{miss} = SVM for NOT-normal prediction
13:	**end if**
14:	Transmitting D_{miss} to Nor_i and S_{sink}
15:	$m = m+1$;
16:	**end while**
17:	**else**
18:	break;
19:	**end if**

5   Experiment and Result Analysis

In order to verify these above discussing algorithms, we construct an IWSN prototype in lab. So in this section, we briefly report our prototyping experiences on an industrial environment system.

Figure 6 illustrate the hardware architecture and communication protocols used in the prototype. The sensor node is basically a temperature and humidity node with relevant sensor and wireless radio module. ZigBee is adopted in communication protocols. The laptop acts a server, while the nodes transmitting the data to server by ZigBee. Meanwhile, the laptop is used to restore and execute SVM prediction and classification algorithms.

In this set of real-life of experiments, we combine four monitoring parameters in an industrial field (Machinery Factory) to test the performance of the proposed algorithms. We collect the temperature from 13:30 to 17:00, while using the SVM algorithm. As Fig. 5 shown, the experimental temperature of using SVM and real-time measuring temperature are given.

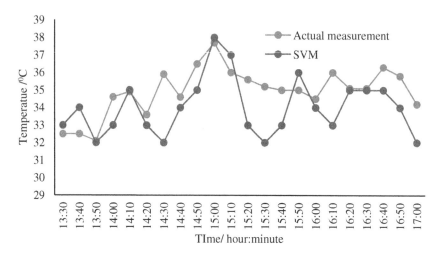

Fig. 5. Using SVM to measure temperature

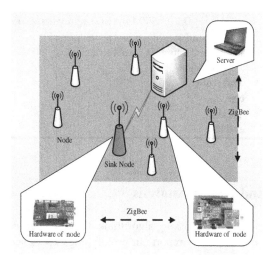

Fig. 6. Implement a real-world industrial WSNs via hardware components

6 Conclusions and Future Work

In this paper, for deal with sensing void holes in IWSNs, we propose a hierarchical support vector machines to recovery data with the objective of providing effective ways for smart sensing industrial environment. The algorithm divides the global data into different levels that means different environmental security levels. Then based on the schemes could recover the missing data of node failure with second layer SVM based on the latest historical records. We attempt to realize a practical system with the prosed algorithm. In the future, based on different other communication protocols such as SMAC, CSMA, and other routing mechanism like LEACH, the proposed way need to verify its feasibility.

Acknowledgments. This work is supported by the Youth Innovation Project of Important Program for college of Guangdong province, China, in 2015 with the No 2015KQNCX228. Meanwhile, this work partly was supported by the colleagues in the Department of Electronic Communication & Software Engineering Nanfang College of Sun Yat-sen University.

References

1. Li, X., Li, D., Wan, J., Vasilakos, A., Lai, C., Wang, S.: A review of industrial wireless networks in the context of industry 4.0. Wirel. Netw. (2015). doi:10.1007/s11276-015-1133-7
2. Zhang, D., Wan, J., Liu, Q., Guan, X., Liang, X.: A taxonomy of agent technologies for ubiquitous computing environments. KSII Trans. Internet Inf. Syst. 6(2), 547–565 (2012)
3. Liu, J., Wang, Q., Wan, J., Xiong, J.: Towards real-time indoor localization in wireless sensor networks. In: Proceedings of the 12th IEEE International Conference on Computer and Information Technology, Chengdu, China, pp. 877–884, October 2012
4. Al Ameen, M., Liu, J., Kwak, K.: Security and privacy issues in wireless sensor networks for healthcare applications. J. Med. Syst. 36(1), 93–101 (2012)
5. Fontana, E., et al.: Sensor network for monitoring the state of pollution of high-voltage insulators via satellite. IEEE Trans. Power Delivery 27(2), 953–962 (2012)
6. Shu, Z., Wan, J., Zhang, D., Li, D.: Cloud-integrated cyber-physical systems for complex industrial applications. ACM/Springer Mobile Netw. Appl. (2015). doi:10.1007/s11036-015-0664-6
7. Yi, J.M., Kang, M.J., Noh, D.K.: SolarCastalia: solar energy harvesting wireless sensor network simulator. Int. J. Distrib. Sens. Netw. 2015, 1–10 (2015)
8. Liang, W., et al.: Survey and experiments of WIA-PA specification of industrial wireless network. Wirel. Commun. Mobile Comput. 11(8), 1197–1212 (2011)
9. Nguyen, K.T., Laurent, M., Oualha, N.: Survey on secure communication protocols for the internet of things. Ad Hoc Netw. 32, 17–31 (2015)
10. Wan, J., Zou, C., Zhou, K., Rongshuang, L., Li, D.: IoT sensing framework with inter-cloud computing capability in vehicular networking. Electron. Commer. Res. 14(3), 389–416 (2014)
11. Wan, J., Zhang, D., Sun, Y., et al.: VCMIA: a novel architecture for integrating vehicular cyber-physical systems and mobile cloud computing. Mobile Netw. Appl. 19(2), 153–160 (2014)
12. Wan, J., Zou, C., Ullah, S., et al.: Cloud-enabled wireless body area networks for pervasive healthcare. IEEE Netw. 5, 56–61 (2013)
13. Wu, J., Yang, S.: SMART: a scan-based movement-assisted sensor deployment method in wireless sensor networks. In: Proceedings of IEEE INFOCOM, pp. 2313–2324, March 2005
14. Zou, Y., Chakrabarty, K.: Sensor deployment and target localization based on virtual forces. In: Proceedings of IEEE INFOCOM, pp. 1293–1303, April 2003
15. Lin, T.-Y., Santoso, H.A., Wu, K.-R.: Global sensor deployment and local coverage - aware recovery schemes for mart environments. IEEE Trans. Mobile Comput. 14(7), 1382–1396 (2015)

Incremental Configuration Update Model and Application in Sponsored Search Advertising

Wei Yuan[1(✉)], Pan Deng[1], Biying Yan[1], Jian Wei Zhang[2],
Qingsong Hua[3], and Jing Tan[4]

[1] Institute of Software Chinese Academy of Sciences, Beijing, China
{814498287,30081046,352957054}@qq.com
[2] Beihang University, Beijing, China
1177926@qq.com
[3] Power System Research Centre, Qingdao University, Qingdao, China
8988596@qq.com
[4] Shenyang Artillery Academy, Shenyang, China
tanj_tanjing@163.com

Abstract. Sponsored search advertising is a significant product contributing impressive revenue to an internet search company. It serves for thousands of partners including owned properties and syndication partners. To boost revenue, configuration should frequently be changed to satisfy every partner's personalized requirements, partners on-boarding or off-boarding, features enabling or disabling, etc. To better address these demands and make them take effect rapidly in system, a model for incremental configuration update and its application in sponsored search are put forward in this paper to do configuration update automatically. The results in application in product system turn out to be encouraging demonstrating its efficiency, helping company reduce costs .

Keywords: Sponsored search advertising · Incremental Configuration Update Model · Data reload

1 Introduction

As one of the largest and fastest growing advertising channels [1], sponsored search is the delivery of relevance-targeted text (image and video ads are also included in large general search engines such as Google.com, Bing.com and Yahoo.com) advertisement as part of the search experience [2]. In Q1 of 2015 [3], 54.6 billion explicit core desktop searches which is free for consumers were conducted in the top 5 American search engines, indicating a robust 26 % growth for the US paid search market [4]. Moreover, advertising expenditures on sponsored search is forecast to grow to more than $110B in 2015. By contrast, other media format (like television, newspaper) advertising spending in the US will encounter a significant decline in growth. Obviously, sponsored search advertising is becoming a key or preferred market promotional tool for more and more enterprises and organizations.

© ICST Institute for Computer Sciences, Social Informatics and Telecommunications Engineering 2016
J. Wan et al. (Eds.): Industrial IoT 2016, LNICST 173, pp. 70–78, 2016.
DOI: 10.1007/978-3-319-44350-8_8

In practice, all major sponsored search players use keyword auctions to allocate display space of advertising material alongside other functionality landing pages. Generally, the keyword auction is based on a pay-per-click model in which advertisers are charged only if their advertisement is clicked by a user [5, 6]. For a specific product or service, advertisers select suitable keywords under some professional or keyword recommendation tools likely to be searched by users in the search page. Meanwhile, advertisers need to provide a bid for each of those keywords indicating the amount of money they would pay for a click. When a query is searched, the engine matches the keywords of all advertisers again the keyword in this query and decides the most eligible ad in an auction which is achieved by a mechanism, namely a generalized second price auction [5].

Along with the crazy growth of sponsored search adverting users, the online serving system is also expanding rapidly to meet amount of personalized demands. Generally, this system usually contains a bunch of specific configuration to manage and serve for hundreds of thousands of customers. Configuration Management (CM) [7] (e.g., configuration version control, changes adoption) becomes more complicated due to continuous customer requirement changes. Configuration need to be updated frequently and most of requirements only impact part of the configuration as usual. This process should also not impact the online system without restarting the engine and come into effect in real time.

In this paper, an incremental configuration update model is abstracted to improve the velocity and efficiency for these frequency changes from practice in Sponsored Search. This model is also applied to automatic configuration update system offering 24 × 7 h reliable service.

The paper is organized in the following fashion. Our scheme proposed for incremental configuration update is detailed in Sect. 2. Section 3 describes the application in Sponsored Search. Some results and analysis are presented and discussed in Sect. 4. Finally, the paper concludes in Sect. 5.

2 Incremental Configuration Update Model

2.1 Elements of This Model

To achieve the practice requirements of pushing the only to-update parts for all kinds of configuration to production as automatically and rapidly as possible with less engineering resources, this paper abstracts the Incremental Configuration Update Model.

Figure 1 shows this model briefly. There are four mainly steps: (1) incremental configuration extractor; (2) quality verification; (3) cross-colo delivery; (4) delta configuration reload. The whole procedure can be described as follows: the requestor provides to-update configuration changes via specific portal which also can cooperate to do some verification; Then, the incremental configuration are derived according to the deriving strategy in step (1) and distribute the results to step (2) for quality verification; After all automatic verification pass and signed by the original requestor, step (3) deploys the incremental configuration to production environment of cross-colo; Finally, step (4) reloads the delta configuration to take effect without any server changes.

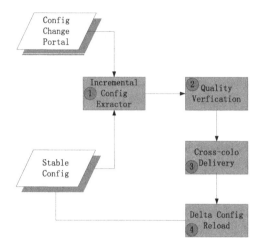

Fig. 1. Incremental configuration update model

2.2 Derive Delta Configuration

There are usually two dimensions for configuration data. One dimension is we called partner, which is a unique value to identify the various advertisers or publishers, and another dimension is we called feature, which is used to describe one special behavior for one or some partners. So in consequence, one partner may include one or more features, and one feature can belong to one or more partners.

In Table 1, "/search/amazon/xml/" represents the partner Amazon and it enables feature 'Sitelink' [10, 11] and partner Pinterest (/search/pinterest/xml/) enables feature 'Sitelink' and 'ImageAds' [12].

Table 1. Partner and feature

Index	Partner	Feature
1	/search/amazon/xml/	Sitelink
2	/search/pinterest/xml/	Sitelink
3	/search/pinterest/xml/	ImageAds

The incremental configuration is usually to express one or more features enabled/disabled for some partners, so the key part of updating incremental configuration is to divide the configuration and compare with the last stable configuration to extract the incremental parts. We can divide the configuration by dimension. Thus, it can be elicited to three methods.

(1) Dividing configuration by partner. In this situation, configuration data are stored in a whole file, and can be divided into many small files, and each of those files is used to indicate one partner configuration. Every partner configuration file contains the whole feature sets.

(2) Dividing configuration by feature. In contrast with the first method, the whole configuration file is divided into many small files according to features. Every feature configuration file contains all the partners who enable this feature.

(3) Dividing configuration by both partner and feature. In this method, each small configuration file only contains the setting for one feature on one partner. Obviously, this method leads to so many file pieces.

But, how to choose the proper dividing method? The goal should be to choose the one which creates minimum (most necessary) incremental configuration for each update. Based on this goal, we should consider following criteria.

a. Compare the size of partner set and feature set. Bigger size leads to more file pieces and impact more for each update.

b. Consider on the pattern for each update. If we immerse ourselves in developing on one new feature, the second method is more proper for each update. If the main activity for each day is to introduce new partner or enable/disable feature for some partners, the first method is a better choice.

2.3 Incremental Configuration Quality Verification

For an automatically updating system, continuous integration and self-verification are critical. The test automation and propagation automation tool can enormously make the verification easier and simpler. Continuous Integration (CI) [8, 9] is a continuous integration test tool and globally used in Internet company. CI job is comprised of abundant basic test cases and can be triggered by every code or configuration change. GDT is automatically propagation tool internally and commonly used for data deployment by service engineer (SE) team for production environment propagation [13, 14]. Once the changes are committed, CI job is triggered automatically. The job consists of following components.

(1) Basic test cases: functional test

(2) Integration test: aim to simulate the incremental configuration launch and verify the incremental configuration and launch. Establish the current production environment in CI and simulate incremental configuration propagation with tool like GDT. And the test queries are created automatically based on the incremental configuration and sent to test environment for integration.

With continuous improvements, the whole model could provide reliable verification for both incremental configuration and launch.

2.4 Cross-Colo Delivery

After the quality of the incremental configuration are verified, the next key component is to deliver the incremental configuration to different data centers. The reason we need cross-colo delivery is that the large-scale system runs in multiple data centers to support BCP or global service.

The key point for this step is to carry the data to the destinations quickly, consistently and under monitoring.

2.5 Configuration Data Reload Automatically

When the incremental configuration arrive at destination machines, the system needs to have the capability to reload it into machine memory and make it come into effect. The reloading should process smoothly, completely, and under monitoring.

(1) Smoothly means the system should not discontinue the service and then switch to serving with the new configuration for new requests.
(2) Completely means the system must reload the whole incremental configuration at once, and if there are some failures happened, the system must cancel the reloaded configuration and rollback to last stable status.
(3) Under monitoring means the whole process should be monitored and alerts should be sent out for both reloading success or not.

3 Application in Sponsored Search

In Sponsored Search engine, there are various configuration data, like partner configuration, system configuration, server configuration, etc. Generally, as partner business and server requirements, the configuration data need to change frequently and update to production as real-time and reliably as possible.

For partner configuration, we have been applying the incremental configuration update model in Sponsored Search. According to production online feedback, it can overwhelming shorten the partner configuration update cycle from one month to daily, even aiming to near real-time in the future; it has reduced engineering resource enormously with only Partners and Account Managers involved in this model and other work are finished by model automatically; The most important point is that it is more reliable because it can verify itself and get verification from requestor.

3.1 Partner Delta Configuration Deriving

In this step, method "Dividing configuration by partner" is used, because the partner size is much bigger than the feature size, and there are more requests to update the partner's existed configuration than developing new feature.

Figure 2 illustrates how incremental partner configuration is derived automatically:

(1) Account Manager will collect requirements from all partners and advertisers, and finish the requirements on config change portal;
(2) Original partner config CI flattens partners configuration changes and hands it over to upcoming CI;
(3) Extractor CI divides the whole partner configuration file into single files according to partners, and compare them with production stable configuration to extract

incremental configuration, then pack it for next step Incremental Configuration Quality Verification.

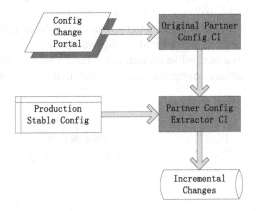

Fig. 2. Partner incremental configuration deriving

3.2 Partner Incremental Configuration Quality Verification

When Verification CI receives incremental configuration, it pushes them into verification runway which is imaged from production environment and send query for this partner configuration changes to verify. After machine verification, it sends result and query sample to Account Manager for launch staging via mail. This step can guarantee changes quality which meets the partners' requirements and system standard. Then, it delivers to next step Cross-Colo Delivery (Fig. 3).

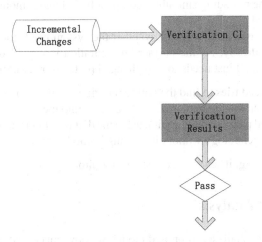

Fig. 3. Partner incremental configuration quality verification

3.3   Partner Incremental Configuration Cross-Colo Delivery

There are amount of tools to support file delivery in cross-colo in different companies. However, it is very crucial to choose proper tools and syncretize them into pipeline.

When the CI gets the successful signal of incremental configuration verification from previous step, it delivers the incremental configuration to different colos. For every launch trigger, it creates a new folder including the incremental configuration and then relink the production partner configuration to soft link to this new folder for following operations (Fig. 4).

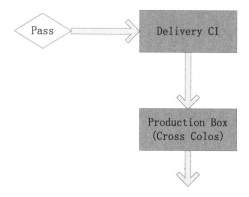

Fig. 4. Partner incremental configuration cross-colo delivery

3.4   Partner Incremental Configuration Reload Automatically

In order to make the reloading smoothly, completely and under monitoring, following process are designed.

A timing thread is created to check whether the system needs to reload the partner configuration periodically. As the partner configuration has divided into single files according to partners, it just needs to find changed partner configuration and reload.

(1)   The timing thread tries to find the partner configuration as per soft link time
(2)   Read the file and load incremental configuration into memory
(3)   Replace existed partner configuration using the delta configuration through memory buffer-switch technology without impacting normal serving

For every reloading, it sends out events for monitor.

4   Results and Analysis

Having described in details our proposed model, we now concentrate on its evaluation. For this purpose, some evaluating indicators have been collected before and after using this model in production launch.

Table 2 shows the application results, indicating the partner configuration to push to production reduce from 24 MB to 80 KB; manpower resource reduces sharply from

8 man/day to 0.5 man/day. The launch cycle shortens to daily and there is no error after using this model for launch.

Table 2. Application results

Item	Before using model	After using model
File size	24 MB	80 KB
Resource	8 man/day	0.5 man/day
Launch cycle	One month	Daily
Error	1	0

Notes: Error represents the number that the changes do not meet requirements during verification.

5 Conclusion

This Incremental Configuration Update Model is extracted based on practice in partner configuration management of Sponsored Search. The application has shown that this model can accelerate the update speed, improve the quality and reduce the involved resource. Of course, this model can be also useful for other large serving systems and industries.

Although it gets some impressive improvement after applying this model in project, there are still some improvements to do in future.

(1) Each step of the model can be improved to reduce latency to reach to the goal that the configuration update can be near real-time from happening to online
(2) Enhance the notification and archiving. Currently, the model only sends a mail after each successful/fail launch and no other records. The function of archiving each change to twiki automatically is a good improvement.

References

1. Yao, S., Mela, C.F.: A dynamic model of sponsored search advertising. Mark. Sci. **30**(3), 447–468 (2011)
2. Fain, D.C., Pederson, J.O.: Sponsored search: a brief history. Bull. Am. Soc. Inform. Sci. Technol. **32**(2), 12–13 (2006)
3. Accessed http://tinyurl.com/nb6cazh, http://tinyurl.com/ph59teu, http://tinyurl.com/nkd3xoe
4. Accessed http://searchengineland.com/report-q1-us-paid-search-growth-strongest-in-3-years-217850
5. Edelman, B., Ostrovsky, M., Schwarz, M.: Internet advertising and the generalized second price auction: selling billions of dollars worth of keywords. Am. Econ. Rev. **97**(1), 242–259 (2007)
6. Graepel, T., Candela, J.Q., Borchert, T., Herbrich, R.: Web-scale Bayesian click-through rate prediction for sponsored search advertising in Microsoft's Bing search engine. In: 27th International Conference on Machine Learning, ICML, pp. 13–20 (2010)
7. Dart, S.: Concepts in configuration management systems. In: Proceedings of the 3rd International Workshop on Software Configuration Management, pp. 1–18 (1991)

8. Accessed https://jenkins-ci.org/
9. Accessed https://en.wikipedia.org/wiki/Continuous_integration
10. Varian, H., Chan, W., Jindal, D., Ranganath, R., Patel, A.: Facilitating the serving of ads having different treatments and/or characteristics, such as text ads and image ads, US 20050251444 A1
11. Cheng, H., van Zwol, R., Azimi, J., Manavoglu, E., Zhang, R., Zhou, Y., Navalpakkam, V.: Multimedia features for click prediction of new ads in display advertising. KDD, 777–785 (2012)
12. Rubens, S., Thomas, D.: Advertising search system and method, US 20060287919 A1
13. Yuan, W., Deng, P., Taleb, T., Wan, J., Bi, C.: An unlicensed taxi identification model based on big data analysis. IEEE Trans. Intell. Transp. Syst. **99**, 1–11 (2015)
14. Wan, J., Lai, C., Mao, S., Villar, E., Mukhopadhyay, S.: Innovative circuit and system design methodologies for green cyber-physical systems. Microprocess. Microsyst. **39**(8), 1231–1233 (2015)

Cloud Computing

CP-Robot: Cloud-Assisted Pillow Robot for Emotion Sensing and Interaction

Min Chen[1], Yujun Ma[1], Yixue Hao[1], Yong Li[2], Di Wu[3], Yin Zhang[4], and Enmin Song[1(✉)]

[1] School of Computer Science and Technology,
Huazhong University of Science and Technology, Wuhan 430074, China
{minchen2012,esong}@hust.edu.cn, yujun.hust@gmail.com,
yixue.epic@gmail.com
[2] Tsinghua National Laboratory for Information Science and Technology,
Department of Electronic Engineering, Tsinghua University, Beijing, China
liyong07@tsinghua.edu.cn
[3] Department of Computer Science, Sun Yat-sen University,
Guangzhou 510006, China
wudi27@mail.sysu.edu.cn
[4] School of Information and Safety Engineering,
Zhongnan University of Economics and Law, Wuhan, China
yinzhang@znufe.edu.cn

Abstract. With the development of the technology such as the Internet of Things, 5G and the Cloud, people pay more attention to their spiritual life, especially emotion sensing and interaction; however, it is still a great challenge to realize the far-end awareness and interaction between people, for the existing far-end interactive system mainly focuses on the voice and video communication, which can hardly meet people's emotional needs. In this paper, we have designed cloud-assisted pillow robot (CP-Robot) for emotion sensing and interaction. First, we use the signals collected from the Smart Clothing, CP-Robot and smart phones to judge the users' moods; then we realize the emotional interaction and comfort between users through the CP-Robot; and finally, we give a specific example about a mother who is on a business trip comforting her son at home through the CP-Robot to prove the feasibility and effectiveness of the system.

Keywords: Emotion sensing · ECG · Smartphone · CP-Robot

1 Introduction

With the development of physical world through various technology advances on Internet of Things (IoT), 5G and clouds, etc., more and more people start to shift their concern on their spiritual life [1–4]. Using cloud-based solutions has been dramatically changing industrial operations in multiple perspectives, such as environmental protection [5], mobility usage [6], and privacy [7]. Though voice and video

© ICST Institute for Computer Sciences, Social Informatics and Telecommunications Engineering 2016
J. Wan et al. (Eds.): Industrial IoT 2016, LNICST 173, pp. 81–93, 2016.
DOI: 10.1007/978-3-319-44350-8_9

communications among people are convenient nowadays, the feeling of face-to-face communications with emotion interaction is still hard to be obtained, especially when people miss their families and friends during their working or sleeping time [8,9]. On the other hand, the future 5G communication systems have higher and higher transmission rate which provide the basis to support the emotion interaction by remote communications [10]. As we know, traditional phone and video call are lack of emotion interaction and body contact, people who need special emotion care may not be satisfied with the experience of verbal based communications [11,12]. For example, people always on business trips or taking some special job, such as sailor, easily feel lonely and need more interactions with family members rather than simple voice communications [13,14]. So it is essential to design a method of enabling communications with more authentic feeling through conventional mobile phone and some new device [15–19].

As one method for human computer interaction, the interaction through robots is attracting more and more attentions [20–23]. In this paper, we specially consider pillow robot for emotion communications due to its intrinsic features of low cost and convenience to carry. In the past, the functions of pillow robot are simple. Usually, a pillow robot is used for entertainment by simulating human's heartbeat by vibrator or body temperature through tiny heater. By comparison, this paper investigates the efficacy of pillow robot for monitoring user's health status and comforting user's emotion. To verify our idea, a new type of pillow robot will be built up [24]. The design goal of such pillow robot is to outperform existing ones in terms of intelligent interaction with user, physiological data collection, integration with cloud system, emotion detection, etc.

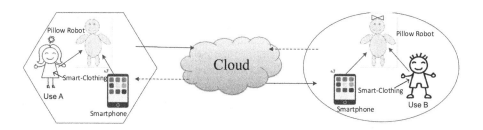

Fig. 1. Illustration of CP-Robot for emotion sensing and interaction.

The novel pillow robot is called as Cloud-assisted Pillow Robot for Emotion Sensing and Interaction (CP-Robot). Compared with traditional pillow robots, the CP-Robot relies on wireless signals for communicating with its user (i.e., CP-Robot owner, denoted as User A) and remote partner (i.e., the subject who holds the other CP-Robot, denoted as User B). Let CP-Robot A and CP-Robot B denote the pillow robots held by User A and User B, respectively. First, the pair of CP-Robots collect the body signals of both User A and User B. When the two users call each other, their CP-Robots work as smart phones. Additionally, users' body signals are transmitted to cloud via CP-Robots for emotion

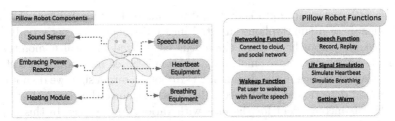

(a) Illustration of functional compo- (b) Illustration of functions of
nents.   CP-Robot.

Fig. 2. Illustration of CP-Robot.

detection. Then, the data associated with User A's emotion is sent to CP-Robot
B. Likewise, CP-Robot A is also aware the emotion status of User B, as shown
in the Fig. 1. Through specially designed functions such as embracing forces,
heartbeats, sounds, and temperatures, CP-Robot A mimics User B's behavior and emotion, while User B can imagine a touchable partner by considering
CP-Robot B as User A. Therefore, CP-Robot brings people a sense of interaction. Inside the pillow, there are several sensors which may have the perception of
different external factors and corresponding feedback devices, including heating
device (simulating body temperature), vibration device (simulating the heartbeat), sound playing device (calling), and air pump (simulating embrace). The
related sensors include the embracing force detection sensor, heartbeat detection sensor, temperature detection sensor, and speech signal detection signal
on human beings. Figure 2 shows the appearance and function modules of the
CP-Robot.

Most of the traditional interactive robots focus on human-robot pair, that is,
realizes the interaction between human and robot [25]. By comparison, CP-Robot
achieves the emotion interaction between two persons geographically separated.
In past years, some work designed remote emotional interaction system based on
smart phone [26]. However, the emotional interaction is only through swinging
the arms to show each other's feelings, without involving more health monitoring and emotional interaction. In the design of CP-Robot, the system not only
enables User A to feel the heartbeat and body temperature of CP-Robot A,
but also provides User A important feedback to sense the remote emotion and
feelings of User B. Actually, in order to realize the functionality of emotion interaction in CR-Robot, two basic functions are also needed: (1) health monitoring
and healthcare; (2) emotion detection. First, the prerequisite of emotion interaction is that the system can detect and recognize user's emotion accurately.
Second, the body signals collected through wearable devices are important data
sources for emotion detection.

Similar with body area networks and various health monitoring systems,
there are three main components for realizing healthcare function of CP-Robot:
(1) the internal sensors hidden in CP-Robot body collects environmental parameters around a user, and sense the user's body signals when he/she hugs the

CP-Robot; (2) the sensor data is delivered to remote cloud via WiFi or through 3G/4G signals of a smart phone inserted into CP-Robot; (3) user's body signals stored in remote healthcare cloud will be utilized for health monitoring and remote healthcare. Since emotion interaction is based on emotion detection, which is a key challenge in the design of CF-Robot. In our design, we do not consider emotion detection through face video or audio, since the retrieval of those information is inconvenient for the user with various limitations. In this paper, we mainly use smart phone and Electrocardiograph (ECG) signals for emotion detection [27,28], Some existing work advocate learning based on multi-modal data can improve the accuracy of emotion detection [29,30]. In this paper, we use the data collected from the smart phone in the daytime and smart clothing at night, make feature extraction to the user's emotional data, and then the Continuous Conditional Random Fields (CCRF) is used to identify the user's emotion from the smart phone and the smart clothing, respectively. At last, we give user's emotion by decision-level fusion.

The main contributions of our research are as follows:

- We have designed CP-Robot system. The two sides can feel the other side's physical and emotional state, so as to realize the real interaction between people through the robot.
- We utilize the data collected by the smart phone and the ECG signal collected by the smart clothing to implement multimodal emotion detection.

The remainder of this article is organized as follows. The emotion sensing and interaction is described in Sect. 2. We set up one practical system platform in Sect. 3. Finally, Sect. 4 concludes this paper.

2   Emotion Sensing and Interaction

In this section, we give a detailed introduction about the emotion sensing and interaction in this system, and the emotion sensing and interaction are divided into 2 situations mainly according to whether the users hug the CP-Robot or not: (1) When the users hug the CP-Robot: if they wear the Smart Clothing, the CP-Robot is able to feel the users physiological signals (such as ECG Signals, temperatures, etc.), and thus realize the emotion sensing; if they do not wear the Smart Clothing, the CP-Robot feels the users' emotions mainly through the signals collected by the smart phones they carry with them; as for the CP-Robot, it can feel the users' hug strength, surrounding environment, etc., to realize the emotional interaction. (2) When the user does not hug the CP-Robot, just similar to the above, the robot feels the user's emotion mainly through the signals collected by the smart phone. When the robot discovers that the users are in bad moods, it can realize the emotional interaction through the pillows. To be specific, first, we collect information through the Smart Clothing, CP-Robot and smart phones; then we have the preprocessing and feature extraction about the data, make use of the characteristics on the Continuous Conditional Random Fields (CCRF) to have the emotion recognition, and give out the users'

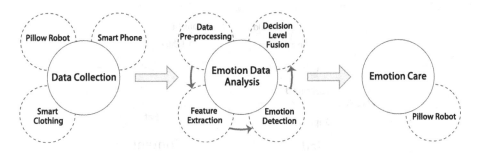

Fig. 3. Illustration of emotion sensing and interaction.

emotions on the basis of the decision-level fusion; finally, we realize the emotion care on the users through the CP-Robot, as shown in the Fig. 3.

2.1   Data Collection

As for the users' data we have collected, we divide it into two kinds: (1) sensing data, namely the data collected by the CP-Robot, the Smart Clothing, and smart phones; (2) labelled data, mainly aiming at tagging people's emotions.

Recognition Data. We make use of the CP-Robot, the Smart Clothing and smart phones the users carry with them to identify the users' emotions: the Smart Clothing is able to collect the users' ECG signals in real time, which are the major defining characteristics to judge the users' emotions; when the users hug the CP-Robot, it can not only feel the strength of the hug, but also realize the functions of the smart phones; the smart phones are able to collect some information such as the users' locations, living habits, etc. in real time. As for the emotion recognition, it is mainly based on the data collected by the Smart Clothing and smart phones. The following part mainly introduces the data collected by the Smart Clothing and smart phones.

- Smart Clothing. The Smart Clothing, a kind of textile products with wearable flexible textile sensors integrated into them, is mainly used to collect the users' ECG signals without making them uncomfortable.
- Smart phone. The smart phones mainly collect users' living habits as well as behavioral data, including their phone call logs, short message logs, application using logs, locations, accelerated speeds, etc., with the data being collected every 10 min.

In addition, the collected data also includes date and time of day, with the date including weekend, weekday and special day, and time of day including morning, afternoon and evening.

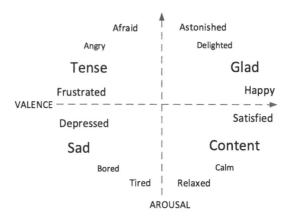

Fig. 4. The Circumplex model of mood: the horizontal axis represents the valence dimension and the vertical axis represents the arousal dimension.

Label Data. As for the tags of the emotion data, we make the users tag their own emotions through phones, with the emotional model used being the dimension affect model. As shown in Fig. 4, it is mainly divided into 2 dimensions: valence and arousal, with the valence being from unpleasant to pleasant, and arousal from calm to active. Also, different valence and arousal are corresponding to different emotions. The users can label on the valence and arousal, and thus we can infer their moods. In this paper, we use several common and representative moods: M = {happy, relaxed, afraid, angry, sad and bored.}.

2.2 Data Preprocessing and Feature Extraction

As for the data collected on the basis of the Smart Clothing and smart phones, we preprocess first and then extract the characteristics.

Data Preprocessing. For the data we have collected, we need to process first, including the cleaning, integration and dimensionality reduction of the data, that is to say, we kick out some missing value, filter the noise in the ECG data, gather together some attributes such as the date, time and use conditions of the phones, and reduce the dimensions of some high dimensional data.

Feature Extraction Based on the Data Collected by the Smart Phone. First, we divide the data collected by the phones into 3 kinds: statistical data, time series data, and text data. (1) The statistical data includes the frequency that the users make phone calls, send short messages and use mobile applications during the data collection period. We can know their characteristics by counting the frequency. (2) The time series data mainly includes users' GPS data and accelerated speed data. Users' location information can be acquired according to the GPS data of their phones, and by using the DBSCAN clustering method,

Table 1. Smartphone feature table

Data style	Data type
Activity level	Static, walking and running
Location	Latitude and longitude coordinates
	User retention time
Phone screen on/off	The time screen on/off
Calls	No. of outgoing calls
	No. of incoming calls
	Average duration of outgoing calls
	Average duration of incoming calls
	No. of missed calls
SMS	No. of receive messages
	No. of sent messages
	The length of the messages
	Content of each SMS
Application	No. of uses of Office Apps
	No. of uses of Maps Apps
	No. of uses of Games Apps
	No. of uses of Chat Apps
	No. of uses of Camera App
	No. of uses of Video/Music Apps
	No. of uses of Internet Apps
Wifi	No. of WiFi signals
	The time use Wifi
SNS	No. of friends
	Content post, repost and comment
	Image post, repost and comment
	Content or Image create time

their visiting locations can be known; then we can judge whether the users are at home, in the office, or out of doors according to their tags; as for the activity data, we record the three-dimensional accelerated speed data x, y, z we have collected as a_x, a_y, a_z, and according to $s = \sqrt{a_x^2 + a_y^2 + a_z^2} - G(Gravity)$ as well as 2 set thresholds $threshold1$, $threshold2$, we can divide them into 3 situations: static ($s < threshold1$), walking ($threshold1 < s < threshold2$) and running ($s > threshold2$). (3) As for the collected text data, we extract the adjectives and nouns in the texts, and turn them into $[-1, -0.4] \cup [0.4, 1]$ by using SentiWordNet, and then the emotional characteristics in the texts can be acquired. We provide the main characteristics of the data collected on the basis of phones in Table 1.

Feature Extraction for the Date Collected by CP-Robot. As for the ECG data we have collected through the Smart Clothing, we use the Convolutional Neural Network (CNN) to extract the characteristics of the continuous time series signals, with the CNN being introduced briefly below,

- *Convolution Layers*: Convolution layer includes a group of neurons to detect time sequence per unit time window or a part of per image, and the size of each neuron decides the area. Each neuron includes and inputs $\mathbf{x}$ with the same number of training weight w, and a deviation parameter θ through an activation function s. In this study, logistic sigmoid function is defined as $s(x) = \frac{1}{1+e^{-x}}$, and the output value is $\mathbf{y}$, which can be described as the following:

$$y_i = s(\mathbf{x} \cdot \mathbf{w}^i + \theta^i), i = 0, 1, \ldots, m. \tag{1}$$

  where m is the number of neurons. Each neuron scans input-layer sequentially, and then achieve feature mapping after neurons go through convolution. The original signals are strengthened and noises are reduced after the neuron goes through a convolution.
- *Pooling Layers*: Once the feature mapping generates, we adopt pooling function to conduct independent sub-sampling. Usually, average value and maximum value are commonly used pooling functions, in this study, we adopt average pooling. The dimensionality of convolution layer is greatly reduced, and overfitting is avoided as well.
- *Convolutional auto-encoder*: Auto-encoder is a kind of unsupervised learning method, whose goal is to learn one representation method for compression and distribution of data set. In this study, we will train all convolution layers by convolving the auto-encoder.

In this paper, we use three-layer convolution and three-layer pool layer to extract the characteristics of the ECG signal.

2.3   Dimensional Emotion Detection

We use Continuous Conditional Random Fields (CCRF) [31] to identify emotions. CCFR is a kind of undirected graphic model, which has preferable results on the tagging and segmentation of the time series [29]. Since we have extracted the characteristics of the time series, we record them as $\mathbf{X} = \{\mathbf{x_1}, \mathbf{x_2}, \cdots, \mathbf{x_n}\}$ to show the input characteristic value, n to show the quantity of the time quantum of a series, and $\mathbf{y} = \{y_1, y_2, \cdots, y_n\}$ to show the corresponding tagged value of the n time series, so we can calculate the conditional probability as:

$$Pr(\mathbf{y}|\mathbf{X}) = \frac{\exp(\Psi)}{\int_{-\infty}^{\infty} \exp(\Psi) dy}. \tag{2}$$

Fig. 5. Function Graph of APP for emotion sensing and interaction CP-Robot System.

where $\int_{-\infty}^{\infty} \exp(\Psi) dy$ makes the conditional probability turn into the normalization equation of 1, and Ψ is a potential function, with its definition being as follows,

$$\Psi = \sum_{i} \sum_{k=1}^{K1} \alpha_k f_k(y_i, \mathbf{X}) + \sum_{i,j} \sum_{k=1}^{K2} \beta_k g_k(y_i, y_j, \mathbf{X}). \tag{3}$$

where α_k and β_k are the parameters of the model, which provide the reliability of f_k and g_k. f_k is vertex characteristic function, which shows the dependency relationship between $\mathbf{x_{i,k}}$ and y_i. $\mathbf{x_{i,k}}$ means the kth element of the vector $\mathbf{x_i}$; g_k is edge feature function, which shows the relationship between the moods y_i and y_j predicted in the time quantum i and j. And f_k and g_k are given by the following formulas:

$$f_k(y_i, \mathbf{X}) = -(y_i - \mathbf{x}_{i,k})^2. \tag{4}$$

$$g_k(y_i, y_j, \mathbf{X}) = -\frac{1}{2} S_{i,j}^{(k)} (y_i - y_j)^2. \tag{5}$$

where $S_{i,j}$ is the similarity measurement, which describes the joint strength between the two peaks in the full connection diagram and has two types of similarities, with the definitions being as follows:

$$S_{i,j}^{(\text{neighbor})} = \begin{cases} 1 & |i-j| = n \\ 0 & \text{otherwise} \end{cases} \tag{6}$$

$$S_{i,j}^{(\text{distance})} = \exp(-\frac{\|\mathbf{x}_i - \mathbf{x}_j\|}{\sigma}). \tag{7}$$

We use the stochastic gradient descent to train, and finally get the tag of the mood y according to the input characteristics.

2.4 Multimodal Data Fusion

When the signals collected by the Smart Clothing and smart phones exist simultaneously, we need to integrate the data. As for the integration problem of the data [32], we give out the integration on the basis of the decision-level, that is to say, we analyze the moods judged according to the mobile phone data and ECG signals together to figure out the users moods, with the specific definition being as follows: we define $y_s^{\mathbf{X}}$, $y_e^{\mathbf{X}}$ and $y_f^{\mathbf{X}}$ to show the moods judged according to the mobile phone data respectively, and as for the moods judged according to the ECG data as well as the both, they can be given out according to the following formulas:

$$y_f^{\mathbf{X}} = \alpha y_s^{\mathbf{X}} + (1 - \alpha)y_e^{\mathbf{X}}. \tag{8}$$

In this experiment, for the sake of simplicity, we mainly use ECG signals ($\alpha = 0.2$) and obtain the users'cores from valence and arousal respectively to judge the users' moods.

3   A Demonstration System for Emotional Interactive

In this section, we have realized the CP-Robot with emotion sensing and interaction which is put forward in this paper. When the users hug the CP-Robot, the mobile phones inset in it will realize the video display between users during the holding time, and in order to make it more convenient for users, we have designed the popular and easy-to-understand cartoon interface. In this system, the mobile phone we have used is Samsung GALAXY Note, which has GPS, three-dimensional accelerations, WiFi and other basic functions. The system is realized on the Android System, with the CP-Robot and Smart Clothing coming from EPIC laboratory (http://epic.hust.edu.cn), and the cloud platform using Inspur SDA30000 [33], for this software is used to comfort the users. During our implementation process, we take the young mother comforting her child as an example, with the scenes being as follows: the mother Rachel is on a business trip in other parts of the country, and the child *Suri* is at home, missing her mother so much and crying sadly. Just as the picture on the left Fig. 5 shows, you can see the photo of *Suri* when having the real-time video chat: he is sad, as shown by the yellow head icon. The pink heartbeat icon means the heart rate, and *Suri* is in unstable mood, with the heartbeat being 121bpm, relatively high; *Suri*s hug strength on the pillow is 1039pa; *Suri*'s temperature is 125.5F; these are *Suri*'s status. As for *Suri*'s location, she is at home, with the movement rate being 0 and the surrounding noise being 65d. The picture on the right is the video photo when the mother is comforting *Suri*: the mother encourages *Suri* to be happy through her facial expressions, as the yellow head icon shows that

Rachel's emotion is happy, with her mood being relatively calm, and heartbeat being 104bpm, relatively normal. In order to comfort *Suri*, Rachel's hug strength on the pillow robot is 1450pa, relatively high. Rachel is in the office far away from *Suri*, with the movement rate being 0 and the surrounding noise being 65d. The experiment exhibits the effectiveness for user Rachel to comfort her child *Suri*'s emotion remotely through CP-Robot.

4   Conclusion

In this paper, we have designed CP-Robot system, which detects a pair of users' emotional status on the basis of the smart phones and the Smart Clothing assisted by the cloud. The Continuous Conditional Random Fields model is used to identify the users' emotion through the time series signals collected from the smart phones and Smart Clothing. In CP-Robot system, a user's mood status is transferred to the CP-Robot of the other person involved, making the pair of users be able to feel the true status of the other, which realizes the emotion sensing and interaction between people over long distance.

Acknowledgement. This work was supported by the National Science Foundation of China under Grant 61370179, Grant 61572220 and Grant 61300224. Prof. Min Chen's work was supported by the International Science and Technology Collaboration Program (2014DFT10070) funded by the China Ministry of Science and Technology (MOST). Prof. Di Wu's work was supported in part by the National Science Foundation of China under Grant 61272397, Grant 61572538, in part by the Guangdong Natural Science Funds for Distinguished Young Scholar under Grant S20120011187.

References

1. Ge, X., Tu, S., Mao, G., Wang, C.-X., Han, T.: 5G ultra-dense cellular networks (2015). arXiv preprint: arXiv:1512.03143
2. Zhou, L., Yang, Z., Rodrigues, J.J., Guizani, M.: Exploring blind online scheduling for mobile cloud multimedia services. IEEE Wirel. Commun. **20**(3), 54–61 (2013)
3. Hossain, M.S.: Cloud-supported cyber-physical localization framework for patients monitoring (2015)
4. Tsai, C.-W., Chiang, M.-C., Ksentini, A., Chen, M.: Metaheuristics algorithm for healthcare: open issues and challenges. Comput. Electr. Eng. (2016)
5. Qiu, M., Ming, Z., Li, J., Gai, K., Zong, Z.: Phase-change memory optimization for green cloud with genetic algorithm. IEEE Trans. Comput. **64**(12), 3528–3540 (2015)
6. Qiu, M., Chen, Z., Ming, Z., Qin, X., Niu, J.: Energy-aware data allocation with hybrid memory for mobile cloud systems (2014)
7. Li, Y., Dai, W., Ming, Z., Qiu, M.: Privacy protection for preventing data over-collection in smart city (2015)
8. Lai, C.-F., Hwang, R.-H., Chao, H.-C., Hassan, M., Alamri, A.: A buffer-aware http live streaming approach for SDN-enabled 5G wireless networks. IEEE Netw. **29**(1), 49–55 (2015)

9. Lin, K., Wang, W., Wang, X., Ji, W., Wan, J.: QoE-driven spectrum assignment for 5G wireless networks using SDR. IEEE Wirel. Commun. **22**(6), 48–55 (2015)
10. Ge, X., Cheng, H., Guizani, M., Han, T.: 5G wireless backhaul networks: challenges and research advances. IEEE Netw. **28**(6), 6–11 (2014)
11. Lai, C.-F., Chao, H.-C., Lai, Y.-X., Wan, J.: Cloud-assisted real-time transrating for http live streaming. IEEE Wirel. Commun. **20**(3), 62–70 (2013)
12. Zhou, L., Wang, H.: Toward blind scheduling in mobile media cloud: fairness, simplicity, and asymptotic optimality. IEEE Trans. Multimedia **15**(4), 735–746 (2013)
13. Zheng, K., Yang, Z., Zhang, K., Chatzimisios, P., Yang, K., Xiang, W.: Big data-driven optimization for mobile networks toward 5G. IEEE Netw. **30**(1), 44–51 (2016)
14. Zheng, K., Hou, L., Meng, H., Zheng, Q., Lu, N., Lei, L.: Soft-defined heterogeneous vehicular network: architecture and challenges (2015). arXiv preprint: arXiv:1510.06579
15. Lai, C.-F., Wang, H., Chao, H.-C., Nan, G.: A network and device aware QOS approach for cloud-based mobile streaming. IEEE Trans. Multimedia **15**(4), 747–757 (2013)
16. Hossain, M.S., Muhammad, G.: Audio-visual emotion recognition using multi-directional regression and ridgelet transform. J. Multimodal User Interfaces, 1–9 (2015)
17. Wang, G., Xiang, W., Pickering, M.: A cross-platform solution for light field based 3D telemedicine. Comput. Methods Programs Biomed. **125**, 103–116 (2015)
18. Hossain, M.S., Muhammad, G., Song, B., Hassan, M.M., Alelaiwi, A., Alamri, A.: Audio-visual emotion-aware cloud gaming framework. IEEE Trans. Circuits Syst. Video Technol. **25**(12), 2105–2118 (2015)
19. Hossain, M.S., Muhammad, G., Alhamid, M.F., Song, B., Al-Mutib, K.: Audio-visual emotion recognition using big data towards 5G. Mobile Networks and Applications, 1–11 (2016)
20. Clavel, C., Callejas, Z.: Sentiment analysis: from opinion mining to human-agent interaction. IEEE Trans. Affect. Comput. **7**, 74–93 (2015)
21. Fortino, G., Galzarano, S., Gravina, R., Li, W.: A framework for collaborative computing and multi-sensor data fusion in body sensor networks. Inf. Fusion **22**, 50–70 (2015)
22. Fortino, G., Di Fatta, G., Pathan, M., Vasilakos, A.V.: Cloud-assisted body area networks: state-of-the-art and future challenges. Wirel. Netw. **20**(7), 1925–1938 (2014)
23. Gravina, R., Fortino, G.: Automatic methods for the detection of accelerative cardiac defense response
24. Chen, M., Song, E., Guo, D.: A novel multi-functional hugtive robot (2013)
25. Han, M.-J., Lin, C.-H., Song, K.-T.: Robotic emotional expression generation based on mood transition and personality model. IEEE Trans. Cybern. **43**(4), 1290–1303 (2013)
26. Saadatian, E., Salafi, T., Samani, H., Lim, Y.D., Nakatsu, R.: An affective telepresence system using smartphone high level sensing and intelligent behavior generation. In: Proceedings of the Second International Conference on Human-Agent Interaction, pp. 75–82. ACM (2014)
27. Chen, M., Hao, Y., Li, Y., Wu, D., Huang, D.: Demo: lives: learning through interactive video and emotion-aware system. In: Proceedings of the 16th ACM International Symposium on Mobile Ad Hoc Networking and Computing, pp. 399–400. ACM (2015)

28. Nardelli, M., Valenza, G., Greco, A., Lanata, A., Scilingo, E.: Recognizing emotions induced by affective sounds through heart rate variability. IEEE Trans. Affect. Comput. **6**, 385–394 (2015)
29. Soleymani, M., Asghari Esfeden, S., Fu, Y., Pantic, M.: Analysis of EEG signals and facial expressions for continuous emotion detection. IEEE Trans. Affect. Comput. (2015)
30. Koelstra, S., Patras, I.: Fusion of facial expressions and EEG for implicit affective tagging. Image Vis. Comput. **31**(2), 164–174 (2013)
31. Baltrusaitis, T., Banda, N., Robinson, P.: Dimensional affect recognition using continuous conditional random fields. In: 2013 10th IEEE International Conference and Workshops on Automatic Face and Gesture Recognition (FG), pp. 1–8. IEEE (2013)
32. Lahat, D., Adali, T., Jutten, C.: Multimodal data fusion: an overview of methods, challenges, and prospects. Proc. IEEE **103**(9), 1449–1477 (2015)
33. Chen, M., Zhang, Y., Zhang, D., Qi, K.: Big Data Inspiration. Huazhong University of Science and Technology Press, Wuhan (2015)

Cloud Robotics: Insight and Outlook

Shenglong Tang[1], Jiafu Wan[1(✉)], Hu Cai[2], and Fulong Chen[3]

[1] School of Mechanical and Automotive Engineering, South China University of Technology,
Guangzhou, China
{tango_scut,jiafuwan_76}@163.com
[2] School of Electrical Engineering and Automation,
Jiangxi University of Science and Technology, Ganzhou, China
caihu_2014@163.com
[3] Anhui Normal Unviersity, Wuhu, China
long005@mail.ahnu.edu.cn

Abstract. With the development of cloud computing, big data and other emerging Technology, the integration of cloud technology and multi-robot system makes it possible to make the multi robot system with high performance and high complexity. This paper briefly describes the concept and development process of the cloud robot and the overall architecture of the cloud robot system. In this paper, the major elements of cloud robot are analyzed from the point of view of big data, cloud computing, open source resources and robot cooperative learning. The key problems to be solved in the current cloud robot system are proposed. Finally, we prospect the future development of the cloud robot.

Keywords: Cloud computing · Big data · Open source · Cloud robot · Internet of things

1 Introduction

The introduction of automation equipment in industrial production in the past few decades has brought great value to the industrial sector. With the development of industrial robots, the programmed robots have reached high level performance in real time, accuracy, robustness and compatibility. At the end of the 1990s, researchers have developed and improved the control of the robot network interface, the robustness of the operation, and "network robotics" [1] appears.

Robotics network refers to a group of robots connected through wired or wireless communication network [2]. Individual robotic in network robotics is regarded as a node. Through sensing data and information shared among nodes, the operators in remote transmit commands data and accept measurement feedback, thus assuring specific operation completed. But like the single robot, robotics network also faces the inherent physical limitations: Due to the limitations of the robot space volume and other factors, there are obvious limitations in computing and storage capacity of individual robot. This leads to the limited capacity of the traditional network robotics when facing high complexity processing. And the performance improvement has obvious limitations as well.

© ICST Institute for Computer Sciences, Social Informatics and Telecommunications Engineering 2016
J. Wan et al. (Eds.): Industrial IoT 2016, LNICST 173, pp. 94–103, 2016.
DOI: 10.1007/978-3-319-44350-8_10

With the development of cloud computing, big data and other emerging Technology, the integration of cloud technology and multi robot system makes it possible to make the multi robot system with high performance and high complexity. In 2009, the European Union project RoboEarth [3], led by Holland Eindhoven University, is establishing the Wide Web for robots World (Robot World Wide Web). RoboEarth can be seen as a large database on the Internet. The emergence of RoboEarth makes the robot share information learn from each other, and finally update the database to achieve a virtuous closed loop.

With the rapid development of the Internet of things and Industrial 4.0 concept put forward [4, 5], at the Humanoids 2010 conference, Professor James Kuffner of Carnegie Mellon University proposed the concept of "cloud robot" for the first time [6], and further elaborated the potential advantages of cloud robots. The cloud robot soon caused extensive discussion and research. Researchers in Singapore for the construction of the DAvinCi [7], Japanese researchers to build business platform Rapyuta [8], development of open source software package ROS [9] (robot operating system) with efforts of willow garage's team have accelerated the development of robotics cloud. In this paper, we will focus on the following points. First, describe the overall structure of the cloud robot, and then analyze several major elements of the cloud robot ecology, analysis of the current key issues to be solved cloud robot. At the end of the article, prospects for the future development of the cloud robot.

2   System Architecture of Cloud Robotics

Network robotics can be seen as transition state between preprogrammed robot to cloud robot [10]. Previously mentioned, Professor Kuffner James proposed a cloud robot, aimed at transferring high complexity of the computing process through communication technology to the cloud platform. This greatly reduces the computational load of the individual robot. Figure 1 describes the main architecture of the cloud robot.

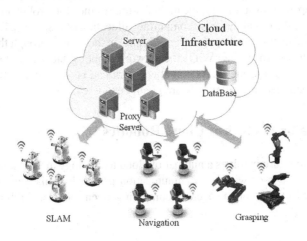

Fig. 1. System architecture of cloud robotics

Cloud robot system is mainly composed of two parts: cloud platform related equipment and bottom facility. The bottom facilities are often all types of mobile robots, unmanned aerial vehicles, machinery and other equipment. Accordingly, the cloud platform is composed of a large number of high-performance servers, proxy servers, massive spatial database and other equipment.

Multi-robot cooperative work, such as SLAM and Navigation [11], is a typical application of cloud robots. Network Robotics' communication mode is used for reference for cloud robotics, and the multi-robot system is composed as a cooperative computing network based on the wireless communication technology. The advantages of cooperative computing networks are the following: (1) The collaborative computing network can gather computing and storage resources, and can dynamically allocate computing and storage resources for specific work requirements. (2) Because of the exchange of information, the machine can be coordinated to make decisions. Unlike network robotic is that when a node's computing ability and storage ability deficiency may lead to large delay time, the node can collaborate around spare node by transferring compute or storage task. For any node which has not connected to the cloud resources, due to the establishment of a link with other robots, they can be connected to the cloud through. The emergence of this mechanism, greatly expands the scope of the task of multi-robot cooperative work and greatly enhance the efficiency of specific work.

Other tasks may not need additional robots but with complexity of the operation, such as Grasping, are other hot research areas. Because the object is unknown, the previous researchers have launch a large number of sensing devices on the grab body [12], trying to improve the accuracy of grasping. But in the actual industrial circumstance, demand of high accuracy and fewer sensors [13] is more in line with the production practice. At the same time, due to the limitation of physical space and material, the equipment and storage equipment are limited. This leads to limited research and development. After the introduction of large data, through uploading a small amount of sensor data and characteristic data, researchers can analyze the match in the cloud, then the characteristic data of the grab motion is downloaded and returned to the mechanical equipment to operate. In summary, the main features of the cloud robot architecture are as follows: (1) In the cloud dynamic computing, allocation of resources, the resource is elastic and on-demand. (2) Cloud robotics' "brain" in the cloud, through the network to obtain the results of processing. While the network robotics' tacks are processed in the body. (3) Because of the fact that computing work can be loaded into the cloud, the robots' load is smaller and battery life are greatly extended.

3   Four Major Factors of Cloud Robotics

Cloud robot's structure involves a number of cloud technology, networking, embedded systems and all kinds of wireless communication protocols. This section we mainly discuss 4 major factors of big data, cloud computing, open source and robot cooperative learning.

3.1 Big Data

Big data as a current hot research field, has been widely used in all kinds of practical research [14]. In the field of robotics, previously mentioned Grasping is a typical example of the use of big data. The significance of large data is not only for its literal meaning, but also for providing an index of pictures, maps and object data of the global database for the terminal (this paper mainly refers to various types of machine equipment). These data include images, videos, maps, sensor networks, and so on. The most typical is the RoboEarth. As the robot database, RoboEarth is attracting a large number of researchers to use and share due to the characteristics of open source. The positive closed loop makes RoboEarth store the massive object data and map data. These important data provide important technical support for the development of robot navigation and grasping.

3.2 Cloud Computing

Cloud computing is originally used for commercial purposes. In recent years, cloud computing has gradually played a role in the field of scientific research [15, 16].

In the robot field, due to the high computing performance of cloud computing, With the computing task uploaded to the cloud, the computational load of computing equipment are liberated to a great extent. For example, In the field of multi-robot operation, cloud computing has greatly accelerated the speed of development of the robot and automation equipment. In order to achieve the robot navigation, Riazuelo et al. [13] used the cloud platform for SLAM, achieving very good results.

However, what we cannot ignore is that the introduction of cloud computing means the upload and download of computing load. Current mainstream communication protocols, such as Zigbee, Bluetooth and Wifi have developed rapidly [17], but the working environment has a great influence on the fluency of wireless communication. The particular example is the interaction of dynamic information which is easy to cause network delay in the process of information transmission. For the applications whose real-time requirement is high, how to use the cloud computing is also a hot research direction.

3.3 Open Source

The spirit of open source promotes the advancement of multi domain technology through the promotion of technological exchange. With the development of cloud technology, the spirit of open source also infiltrates into the cloud robot field. Among them, the most representative is ROS and RoboEarth.

3.3.1 Robot Operating System (ROS)

Robot Operating System (ROS) is a well-known open source, which aims to improve the code reuse rate and development efficiency of the robot operating system.

There are two major parts to mark ROS: system maintenance and distribution.

(1) Main: the core part, designed by Garage Willow and some developers, to provide maintenance. It provides some of the basic tools for distributed computing, as well as the entire ROS of the core part of the program.

(2) Universe: the global scope of the code, Maintained by international ROS community organization. It is a library of code, such as code for OpenCv, PCL. In general, algorithms, frameworks and hardware drivers consist the Universe (Fig. 2).

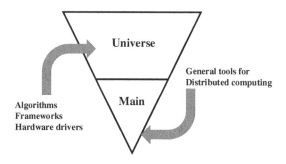

Fig. 2. ROS file architecture

The concept of computation graph is a kind of point to point network form for ROS processing data. When the program runs, all the processes and their data processing, will be displayed through a point to point network. This level mainly includes several important concepts: node, message, subject, service.

Node is a process that performs an operation task. When many nodes are running at the same time, it is easy to draw the point to point communication into a graph. The nodes communicate by sending messages or service. Messages are in a publish / subscribe manner. A node can publish a message in a given topic. A node is concerned with a particular type of data via subcribe on a particular topic. As for service, when the other nodes send request data to the service node, the service node will response. Figure 3 describes the two communication methods between two nodes through the message and service mode.

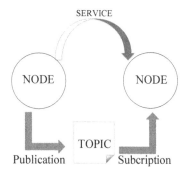

Fig. 3. Point to point communication mode

ROS has numerous nodes, messages, services, tools, and library files that require an effective structure to manage the code. In the ROS file system level, there are a number of important concepts: package, stack. ROS's software is organized in a package. Package containing nodes, ROS dependency library, configuration files, third party software, etc. The objective of the package is to provide a structure that is easy to use in order to facilitate the reuse of the software. The corresponding stack is a collection of packages, which provides a complete set of features. The community level concept of ROS is a code release of a form of expression, and Fig. 4 describes the community level architecture.

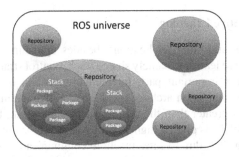

Fig. 4. Community level architecture

3.3.2   RoboEarth

RoboEarth is a web site dedicated to the robot service. It is a huge network database system, the robot here can share information, learn from each other's behavior and environment.

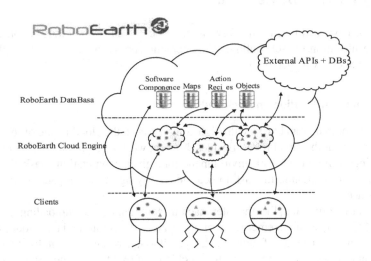

Fig. 5. Architecture of RoboEarth

The overall architecture of the RoboEarth is shown in Fig. 5. In general, RoboEarth is divided into Clients layer, Cloud Engine layer, Database layer. The computing tasks will be transferred to cloud with a unified data format through the network, which makes all the information for uploading of any robot the same. After the unified data format, computing tasks will be transferred to Engine Cloud for cloud computing. Engine Cloud sometimes need to interact with the upper Database data, and finally return the results to the underlying robot. Some applications such as Grasping may directly access the upper layer of the Database for model matching, then the feature data is returned to the underlying robot, in order to carry out further operations.

3.4 Robot Cooperative Learning

In SLAM, navigation and other applications, besides uploading computing tasks to cloud, robot cooperative learning, namely shared track, control strategy and other information is also another important part. Inter communication is typical machine to machine (M2M) communication architecture. Collaborative learning requires that the global task priority is greater than the priority of the node task, the dynamic packet interaction and control. This gives a higher requirement for the control algorithm.

It is worth noting that, due to inter machine communication (M2M) usually use routing proactive protocol and routing Ad-hoc protocol [18]. In the unfamiliar environment network robots may choose proactive routing protocol to open up the path, proactive routing protocol contains a periodic packet switching and routing table updates, which makes the consumption of computing resources and memory resources particularly huge. While according to Ad-hoc protocol, to communicate with the target node, the source node needs to establish the routing table, lead to a more serious delay, this situation is particularly serious in the dynamic network topology.

4 Key Issues to Be Resolved

This chapter we will discuss the key issues to be solved of the cloud robot, namely Resource allocation and scheduling, Data Interaction between robot and cloud platform and Cloud Security respectively.

4.1 Resource Allocation and Scheduling

Uploading computational tasks with high complexity to the cloud is one of the biggest characteristics of the cloud robot. Besides different working equipment, different interface settings, different network environment, for a given computational task, the choice of uploading or assigning the task to near nodes has significant impact on the overall performance.

It is worth noting that, like the inter machine communications, uploading task to the cloud (M2C) may result in emergence of delay. New algorithms and techniques need to face the real time change of network delay. Although wireless technology has made significant progress, once the connection problems of robot and cloud services happen, serious delay is almost inevitable. Therefore, when designing a new algorithm, we need to design a load distribution algorithm with "any time" characteristic. Once it is found that the task that cannot be properly uploaded to the cloud needs to be uploaded to the cloud, mechanism of dynamic allocation of computing tasks should be activated, thereby controlling the delay time.

4.2   Data Interaction Between Robot and Cloud Platform

The most significant feature of cloud robot is data interaction between robot and cloud platform. The output data structure differs greatly among manufacturers. In fact, different models of product from even the same manufacturer may result in output data structure of great difference. The diversity of data structure presents a high requirement for the compatibility of the cloud input interface. To solve this problem, the current mainstream cloud platforms often provide multiple interfaces for numerous formats of data.

However, due to the limited number of interfaces, the data to be uploaded must be properly preprocessed. How the data format needs to be translated greatly affects the robustness and real-time performance of the data exchange. Additionally due to the invariance of the cloud device, cloud devices can only handle and store data of specific structure. This means that the input interfaces need to transform the corresponding data structure into a unified format. Lastly at the output part, when the upload data processing is completed, the ready-to-output data must also be transformed to specific formats. The output data needs to consider the compatibility of the underlying devices and the real-time requirements of the entire data exchange.

4.3   Cloud Security

The introduction of cloud technology has greatly expanded the complexity of multi robot operation. But at the same time it also introduces new technical challenges: the privacy and security issues brought by the cloud technology.

These hidden dangers contain the data generated by the computing devices and sensors in the work of the cloud robot. Commercial science and technology has appeared serious data leakage incidents, especially in the upload to the cloud photos, video [19]. In scientific research and industrial practice, key data stored in the cloud may be far from the hacker to steal, leading to the loss of key data. In order to eliminate these hidden dangers, we must establish the relevant management rules and legal provisions [20, 21]. In addition to the technical level, the establishment of a unified standard and prevention mechanisms as the technical support is also an important prerequisite for the healthy development of the field of cloud robotics.

5   Conclusions

This paper briefly describes the origin and development of the cloud robot, from the allocation of resource, communication mode, and analyzed of the cloud robot communication architecture. Four major elements of the cloud robot are summarized, and the key technologies to solve the development of the cloud robot are analyzed. Cloud robot, sharing computing resources between each other and the cloud platform, pushes multi-robot system and collaborative learning to new height. By sending the computational load of high complexity to the cloud platform, the cloud robot no longer requires high performance of onboard equipment, greatly reducing the cost of the application of multi robot system. With the development of cloud computing, big data and other fields, the cloud robot in SLAM, Grasping, Navigation and other applications will achieve better performance.

References

1. New York Times. http://www.nytimes.com/2014/08/18/technology/for-big-data-scientists-hurdle-to-insights-is-janitor-work.html
2. IEEE Society of Robotics and Automation's Technical Committee on: Networked Robots. http://www-users.cs.umn.edu/~isler/tc/
3. Riazuelo, L., Tenorth, M., Di Marco, D., et al.: RoboEarth semantic mapping: a cloud enabled knowledge-based approach. IEEE Trans. Autom. Sci. Eng. **12**, 432–443 (2015)
4. Wang, S., Wan, J., Li, D., Zhang, C.: Implementing smart factory of industrie 4.0: an outlook. Int. J. Distrib. Sens. Netw. **2016**, 10 (2016)
5. Wang, S., Wan, J., Zhang, D., Li, D., Zhang, C.: Towards smart factory for industry 4.0: a self-organized multi-agent system with big data based feedback and coordination. Elsevier. Comput. Netw. **101**, 158–168 (2016)
6. Kuffner, J.J.: Cloud-enabled robots. In: Proceedings of IEEE-RAS International Conference on Humanoid Robotics (2010)
7. Arumugam, R., Enti, V.R., et al.: DAvinCi: a cloud computing framework for service robots. In: IEEE International Conference on Robotics and Automation (ICRA), pp. 3084–3089 (2010)
8. Mohanarajah, G., et al.: Rapyuta: a cloud robotics platform. IEEE Trans. Autom. Sci. Eng. **12**, 481–493 (2015)
9. Morgan, Q., et al.: ROS: an open-source robot operating system. In: IEEE International Conference on Robotics and Automation (ICRA) Workshop on Open Source Software, vol. 3(2), p. 5 (2009)
10. Hu, G., Tay, W.P., Wen, Y.: Cloud robotics: architecture, challenges and applications. IEEE Netw. **26**, 21–28 (2012)
11. Rekleitis, I.M., Dudek, G., Milios, E.E.: Multi-robot cooperative localization: a study of trade-offs between efficiency and accuracy. In: IEEE/RSJ International Conference on Intelligent Robots and Systems 2002, vol. 3, pp. 2690–2695 (2002)
12. Felip, J., Morales, A.: Robust sensor-based grasp primitive for a three-finger robot hand. In: IEEE/RSJ International Conference on Intelligent Robots and Systems, pp. 1811–1816 (2009)
13. Kehoe, B., Warrier, D., Patil, S., Goldberg, K.: Cloud-based grasp analysis and planning for toleranced parts using parallelized Monte Carlo sampling. IEEE Trans. Autom. Sci. Eng. **12**, 455–470 (2015)
14. Yuan, W., Deng, P., Taleb, T., Wan, J., Bi, C.: An unlicensed taxi identification model based on big data analysis. IEEE Trans. Intell. Transp. Syst. **17**, 1703–1713 (2015)
15. Shu, Z., Wan, J., Zhang, D., Li, D.: Cloud-integrated cyber-physical systems for complex industrial applications. Mobile Netw. Appl., 1–14 (2015). ACM/Springer

16. Wan, J., Zhang, D., Zhao, S., Yang, L., Lloret, J.: Context-aware vehicular cyber-physical systems with cloud support: architecture, challenges, and solutions. IEEE Commun. Mag. **52**, 106–113 (2014)
17. Li, X., Li, D., Wan, J., Vasilakos, A.V., Lai, C.F., Wang, S.: A review of industrial wireless networks in the context of industry 4.0. Wirel. Netw., 1–19 (2015)
18. Royer, E.M., Toh, C.K.: A review of current routing protocols for ad hoc mobile wireless networks. IEEE Pers. Commun. **6**, 46–55 (1999)
19. Ren, K., Wang, C., Wang, Q.: Security challenges for the public cloud. IEEE Internet Comput. **1**, 69–73 (2012)
20. Lin, P., Abney, K., Bekey, G.A.: Robot ethics: the ethical and social implications of robotics. MIT Press, Cambridge (2011)
21. Proia, A.A., Simshaw, D., Hauser, K.: Consumer cloud robotics and the fair information practice principles: recognizing the challenges and opportunities ahead. Soc. Sci. Res. Netw. 145 (2015)

Research of Construction and Application of Cloud Storage in the Environment of Industry 4.0

Kaifeng Geng[(⊠)] and Li Liu

Software School, Computer Network Center, Nanyang Institute of Technology,
Nanyang, 473000 Henan, China
gkf8605@126.com

Abstract. With the geometric growth of data generated from complex system test, experiment and condition monitoring, big data has become the hotspot of Industry 4.0 era. How to meet market demand with efficiency, comprehensiveness and low cost through collection and analysis of data is a problem to be solved. We put forward industrial cloud storage model based on Hadoop on the basis of relevant researches and theories of Industry 4.0, big data Hadoop and so on, and then implement and evaluate each module. It performs well in reliability and expandability through test, providing challenges of big data storage in Industry 4.0 era with effective solution.

Keywords: Industry 4.0 · Hadoop · Cloud storage · Big data

1 Introduction

With the deep merging of informatization and industrialization, information technology has permeated into all aspects of the industrial chain in the field of industry. Bar codes, two-dimensional codes, RFID, Sensors, industrial automation systems, Internet of things and other technologies are widely applied in industry. Especially with the wide application of Internet and Internet of things, industry has entered the new developing stage of internet industry and enterprises have more rich data. In order to make use of data, enterprises need the capacity to support multiple types of information, infrastructures for the store of big data and the capacity of fast and accurate analysis of information after storage.

There is a large amount of data to be collected and processed produced by industrial equipments, the scale of the big data in the data set is from dozens of TB to lots of PB and these data are mostly unstructured in the data type. Moreover, operation of the production line with high speed requires more real-time data, so using traditional data storage schemes to complete the collection and processing of data within an acceptable period of time will be very difficult. With the increasing number of advanced devices and equipments, a large number of operating data goes online, which heralds the coming of Industry 4.0 [1].

To process these high-dimensional data reflecting product information and equipment information, there need prominent computing and storage capacity. Though computers' computing capacity is in improvement, it is still far from the requirement of

© ICST Institute for Computer Sciences, Social Informatics and Telecommunications Engineering 2016
J. Wan et al. (Eds.): Industrial IoT 2016, LNICST 173, pp. 104–113, 2016.
DOI: 10.1007/978-3-319-44350-8_11

processing such enormous data. In addition, in a typical big data storage system, the concurrency of the reading and writing of a variety of data is high, so the restriction of the storage and processing capacity of the central server will inevitably lead to a single point of failure, which blocks the development of the fuse of informatization and industrialization as well as the upgrade and transformation of the whole industry. The birth of cloud computing technology brings dawn to problems such as data processing, storage and so on.

2 Related Concepts of Industry 4.0, Big Data and Hadoop

2.1 Industry 4.0

The research project, "Industry 4.0", was firstly put forward in German and has risen as a national strategy. "Industry 4.0" aims at creating an individualized and digital production model of products and services with high flexibility [2]. The core meaning of Industry 4.0 times is information technology, we can realize the effective and rapid docking of two sides based on the demand and supply of data analysis, can reduce the cost of spending and achieve the directional and customized production through the Internet [3].

2.2 Big Data

Big data is a data set whose scale is much larger than the capable scale of traditional database software tools in acquirement, storage, management and analysis of data. Nowadays, sensors, GPS systems, Internet of things and social networks are creating a new flow of data and big data has become a symbol or a characteristic of Internet in current developing stage. Data itself contains value, so it must flow seamlessly and securely to people when making decisions or action and offer basis for decisions at any time. Under the backdrop of technical innovation represented by the cloud computing, these data difficult to be collected and used in the past are getting easier. And with the constant innovation of all industries, big data will create more value for human beings step by step [4].

2.3 Hadoop Technology

Hadoop [5] Framework is an open source project of Apache Foundation, which is a software framework capable of distributed processing of a large amount of data and offers easy programming interface. It is a platform of cloud computing where programmers can easily develop and handle a huge amount of data. MapReduce and HDFS are the core design of the framework. Hadoop cluster is a specific type of cluster designed specifically for the storage and analysis of massive unstructured data. In essence, it is a kind of computing cluster, which will allocate the data analysis work into multiple cluster nodes, so that the data can be processed in parallel.

MapReduce is a programming model for processing and generating big data sets. The working process of MapReduce is divided into two stages: map stage and reduce stage, among which, "map" is to divide a task into multiple tasks, while "reduce" is to summarize the processing result of multiple divided tasks and then get the final result [6]. The user-defined Map function is used to process a data set based on key/value pair and output the intermediate data set based on key/value, while Reduce function is used to merge all intermediate value with the same intermediate key.

Using the ideas of functional programming language Lisp, MapReduce defines the following two abstract programming interfaces, Map and Reduce, which will be implementated by programming [7]:

map: (k1; v1) → [(k2; v2)]

Input: data presented by key/value pair (k1, v1)

Processing: Document data record (such as a line in a text file, or data in the table rows) will be transfered into the map function in the form of "key/value pair; The map function will deal with these key/value pairs and output intermediate results in the form of another key/value pair [(k2, v2)].

Output: intermediate data presented by [(k2, v2)]

reduce: (k2; [v2]) → [(k3; v3)]

Input: the key/value pair [(k2, v2)] will be merging processed which is output by map, different value under the same primary key will be combined to a list (v2), so the input of reduce is (k2; [v2]);

Processing: the incoming intermediate results will be further processed or listed in some sort, and generate the final output (k3; v3).

Output: the final output [(k3; v3)]

HDFS is a distributed file system designed for storing large files in the mode of streaming data access which can not only customize the block size of store files (the default is 64M), but also customize the number of copies and the security level with high fault-tolerance and security [8].

3   The Influence of the Cloud Computing to Industry 4.0

With the cloud storage getting cheaper and the cloud processing getting more powerful, the cloud becomes the best choice for the storage and analysis of data collected by enterprises. The innate characteristics of the cloud computing, that is, cheap storage and good performance computing make it better than other traditional technologies serving the industry.

3.1   Low Requirements of the Cloud Computing for the Configuration of the Client and the Server, and Low Cost

Hadoop cluster is relatively cheap, there are two main reasons. Its required software is open source, so it can reduce the cost. In fact, you can free download Apache Hadoop distribution. At the same time, the Hadoop cluster controls the costs by supporting commercial hardware. So you don't have to buy the server hardware to build a powerful

Hadoop cluster [9]. One of the core concepts of the cloud computing is to reduce the processing load of the user terminal through constant improvement of the processing capacity of the "cloud". Clients only need to input and output, and all other functions such as computing, storage and processing are managed by the "cloud". Users only need to order relevant services of the "cloud" according to their own needs. In addition, the storage equipments of the "cloud" can be cheap PCs, even old computers. Compared with the single professional storage equipments with large volume, the "cloud" has larger storage capacity and lower storage cost, and can realize dynamic upgrade and extensions according to the demand [10].

3.2 Cloud Computing Can Offer Massive Computing and Storage Capacity, Has a Great Deal of Extensibility

Like any other type of data, an important problem faced by big data analysis is also the increasing amount of data. And the biggest advantage of big data is that it can realize real-time or near real-time analysis and process. Hadoop cluster parallel processing capabilities can significantly improve the analysis speed, but with the increase of the amount of data to analysis, the cluster's capacity is likely to be affected. But thankfully, by adding additional cluster nodes can effectively extend the cluster.

The cloud computing can gather resources such as memories, hard drives and CPUs of all nodes into a giant virtual cooperative working pool of resources, providing storage and computing services for the outside together. With the increase of nodes, the capacity of storage and computing can unlimitedly increase.

3.3 Cloud Computing Can Provide Storage with High Reliability and Security

Data collected in all parts of the industry such as production and sales will be stored multiple service nodes of the cloud with multiple copies. Data stored in the cloud will not be affected even if accidentally deleted. And there is no need of fearing virus invasion and the data loss caused by hardware damage.

4 Industry Cloud Storage Model Construction Based on Hadoop Technology

Hadoop is a tool which can realize data storage expansion by using standard hardware and can distribute data among many low-cost computers. After data distribution follows the difficulties of data location and handling which can be solved by Map Reduce. Map Reduce provides a framework, and data in a cluster are parallel processed among many nodes. It is allowed to map the processing to many location data and cut similar data elements to a single result [11].

Aimed at challenges faced by the construction of big data storage system at the present, on the basis of the advantages of the Hadoop technology, we put forward the industrial cloud storage model based on the Hadoop platform, which can effectively solve problems such as processing capacity limitation, storage capacity limitation and single point failure. In the cloud computing technology, data collected by sensing equipments, bar codes, two-dimensional codes, RFID and so on are provided in the form of saas (software-as-a-service) in the cloud, terminals are responsible for collecting data and sending data to cloud applications, and present massive intuitive data as well as statistic results in all parts of the industrial production, These data are generally widely distributed and unstructured, but Hadoop is very suitable for this kind of data because the work principle of Hadoop is that the data is split into slices, each "slice" will be analyzed by assigning to a specific cluster nodes. The data distribution is not have to uniform, for each shard is processed separately on each independence cluster nodes.

which can not only provide the independency of the platform, but also reduce the possibility of problems caused by loading central server and the single point fault. Meanwhile, using MapReduce model can also avoid single fault in the calculation of lots of high dimensional data. This framework showed in Fig. 1 is divided into the front end and the back end from top to bottom.

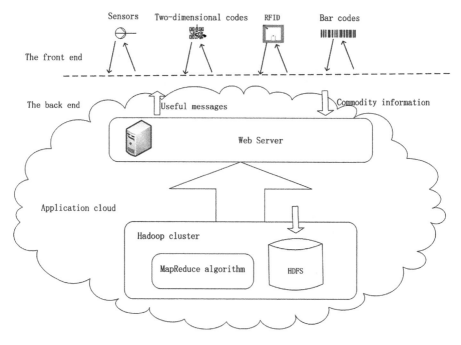

Fig. 1. Industry cloud storage model based on Hadoop technology

4.1 The Front End

The front end consists of the sensor, the bar code, the QR code, RFID and other mobile devices, which is used for collecting data and presenting optimized information. Signals collected from all parts of the industrial production are uploaded to the back end through this application for the further processing, and are presented in the appropriate form.

This information can provides basis for decision making, bringing deep knowledge of the industry and competitive advantages.

4.2 The Back End

The back end is the core of this model, mainly including three modules such as the WEB server, MapReduce algorithm and HDFS cloud storage. WEB server is responsible for the communication between the Hadoop cluster and the WEB interface as well as receiving sensor signals from the front end and displaying all kinds of optimized data. A Hadoop cluster with multiple nodes is responsible for solving parallel processing tasks. Each node has a copy of MapReduce java program and MapReduce is responsible for the large-scale parallel computing of industrial big data according to a certain algorithm. As a distributed file system, HDFS is used to store big data generated in all parts and provides high throughput for accessing application data, ensuring seamless transfer of data among servers and improving the reliability of the whole system.

5   System Implementation

Due to the complex structure and powerful function of a whole industrial cloud storage system, we only implement the cloud environment, MapReduce and HDFS cloud storage. In order to make users' operation convenient, a WEB interface is needed to be developed to upload collected data and display processing results. There need two java scripts for the interaction between the front-end and the back-end: one is for uploading data files of sensing signals to HDFS; the other is for accessing the output files of Reducer, and getting the maximum key/value pair through comparing there key/value pairs whose value can reflect the real condition of the products. Key codes of the uploaded Java script are as follows:

```
String src = "e://degree.txt";
String dst = "hdfs://210.42.241.66:9000/user/Hadoop/degree.txt";
InputStream in = new BufferedInputStream(new FileInputStream(src));
FileSystem fs = FileSystem.get(URI.create(dst), new Configuration(););
OutputStream out = fs.create(new Path(dst), new Progressable() { });
IOUtils.copyBytes(in, out, 4096, true)
```

5.1   Cloud Environment

Cloud environment plays a vital role in the system architecture, providing the whole system with massive storage and computing capacity. In this experiment, we use Hadoop cluster built by 3 PCs as the cloud, among which one serves as the master node, the other three serve as slave nodes. Hadoop version is 0.21.0, JDK version is 1.6 and the operating system is Red Hat Linux5.4. Details of the installation and configuration process need not be repeated here, and nodes' state information after load balance is show in Fig. 2.

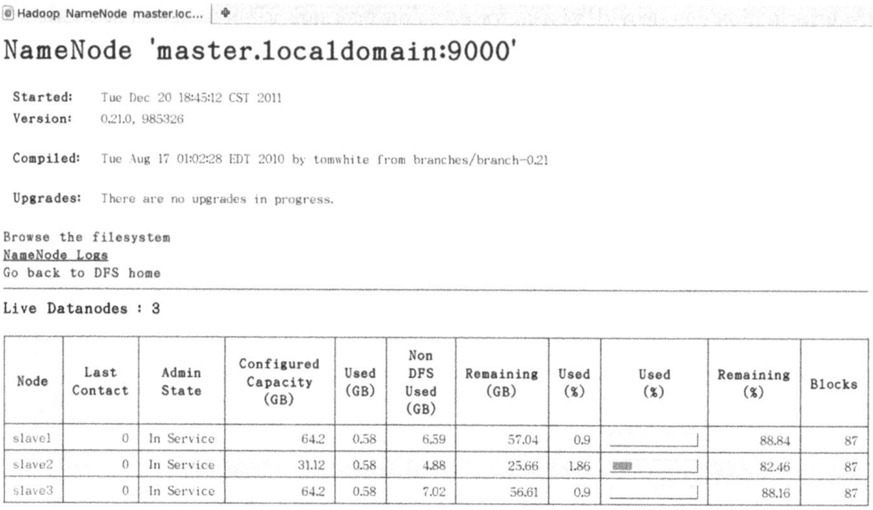

@ Hadoop NameNode master.loc... | ◆

NameNode 'master.localdomain:9000'

Started: Tue Dec 20 18:45:12 CST 2011
Version: 0.21.0, 985326

Compiled: Tue Aug 17 01:02:28 EDT 2010 by tomwhite from branches/branch-0.21

Upgrades: There are no upgrades in progress.

Browse the filesystem
NameNode Logs
Go back to DFS home

Live Datanodes : 3

Node	Last Contact	Admin State	Configured Capacity (GB)	Used (GB)	Non DFS Used (GB)	Remaining (GB)	Used (%)	Used (%)	Remaining (%)	Blocks
slave1	0	In Service	64.2	0.58	6.59	57.04	0.9		88.84	87
slave2	0	In Service	31.12	0.58	4.88	25.66	1.86	▦	82.46	87
slave3	0	In Service	64.2	0.58	7.02	56.61	0.9		88.16	87

Fig. 2. State information of Hadoop cluster nodes

5.2 MapReduce Implementation

The difficulty of industrial data management is not limited to the quantity of information. As data have different formats and sources, there exist problems of diversity and complexity of the data. There often exists processes information "island" which must be merged, stored and analyzed to obtain meaningful values.

In the MapReduce model, each master has a JobTracker (JobTracker is a master service. After the startup of software, JobTracker receives job, and is responsible for scheduling each sub-task of the job to run on TaskTracker, monitoring them and restarting the task if discovering failed task. In general, JobTracker should be deployed in a single machine). Each salve has a TaskTracker (TaskTracker is a slaver service running at multiple nodes. TaskTracker initiatively communicates with JobTracker, receives tasks and is responsible for the direct implementation of each task) [12].

The implementation of MapReduce includes the map stage and the reduce stage: the map stage extracts the characteristics of each line's data in the matrix (the matrix represents the collected data) and gets the evaluation results. Among them, each line in the matrix represents the signal s from an electrode. The inputting of Mapper is a key/value pair. Key is a matrix file and the inputting of Mapper is to generate a new key/value pair, for example, (Good, 1), (Neutral, 2), or (Bad, 4). The reduce stage processes the unique key from the map and then obtains three key/value pairs which represent the amount of Good, Neutral and Bad keys among the totality. For example, key/value pairs, (Good, 2), (Neutral, 2), (Bad, 1), (Bad, 4) and (Good, 1) generate new key/value pairs (Good, 3), (Neutral, 2) and (Bad, 5) after Reducer's processing.

5.3   Implementation of HDFS Cloud Storage

In the framework of HDFS cloud storage, master has a NameNode and slave has a DataNode. NameNode is a central server, responsible for managing the namespace of the file system and the access of the client to files. DataNote is generally responsible for managing the storage of its nodes. HDFS exposes the name space of the file system and users can store data in it in the form of file. Seeing from the inside, a file is actually split into one or more data blocks which are stored in a set of DataNote. NameNode implements the operation of the file system's name space, as Fig. 3 shows.

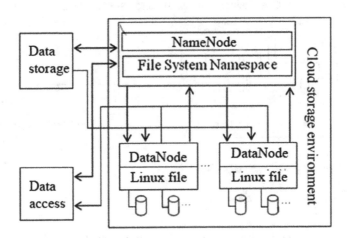

Fig. 3. HDFS cloud storage architecture diagram

6   Evaluation and Testing

We use a large data set to evaluate the system and design two different tests for the objective test performance: load testing and performance testing.

6.1   Specific Environment Configuration

First set of the test:

The configuration of a single computer is as follows:

Platform: Red Hat Enterprise Linux 5 update 4; Processor: 32 Bits, 3.0 GHz Intel Dual-core; memory: 4 GB; drive: 500G. Network card: 100M full duplex network connection.

Second set of the test:

The configuration of the four-node Dadoop computer cluster is as follows (configuration is much lower than the first set of the test):

Platform: Red Hat Enterprise Linux 5 update 4; Processor: 32 Bits, 1.8 GHz Intel Dual-core; memory: 2 GB; Drive: 500G; Network card: 100M full duplex network connection.

6.2   Load Testing

Load testing is to test the scalability and reliability of the system. In order to simulate the loading environment of the system, we use Neoload to record user context and simulate 500 virtual users. Ganglia monitor is used to monitor the average utilization of the node CPU. As Fig. 4 shows, after the test, when there are 500 concurrent users, the utilization of the system CPU is less than 5 %, which shows this system is very energy-efficient.

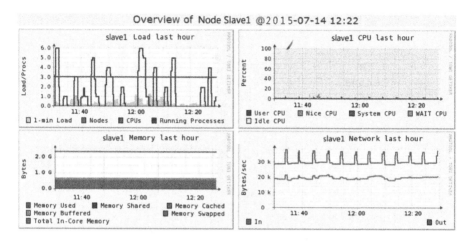

Fig. 4.   Ganglia monitors the status of Slave1 node

6.3   Performance Test

Comparison method is used to evaluate and test the system's overall performance. The first test: write a normal Java application as a skill base for reference in the test and evaluation, which is running on a local machine to calculate the baseline performance; the second test: the same algorithm runs on the Hadoop cluster with 4 nodes to implement Java MapReduce.

When running MapReduce algorithm program, we choose two different users with different amount to concurrently access the system. The first test: simulate 5 virtual users, after the test, estimate the baseline performance of the first test is 0.8 s and the baseline performance of the second test is 5 s. The experiment result shows: Local Java applications are much faster than applications running on the Hadoop cluster. The second test: simulate 500 virtual users, after the test, estimate the baseline performance of the first test is 61 s and the baseline performance of the second test is 22 s. The experiment result shows: Java applications running on the Hadoop cluster are much faster than local applications.

Through above two experiments we can conclude that: in the case of small-scale users, local ordinary Java program responses obviously faster, and compared with traditional Java programs, using HDFS and MapReduce has no advantage. However, in the case of large-scale users, because MapReduce is much potential in processing large files (TB level), it distributes the large-scale operation of the data set to each node on the network for implementation. In general, data amount generated in the environment of Industry 4.0 is relatively large, therefore the proposed system has reasonable structure and good performance [13].

7   Conclusion

Faced with the increasing large amount of data under the environment of Industry 4.0, only using appropriate tools for prediction and analysis can the large amount of disorderly data be processed into usable information. The proposed industrial cloud storage model based on Hadoop technology has good performance through the test which can provide the data analysis and processing of the industry with efficient help and explain some certain uncertainty as well as predict problems that may occur so as to help companies to make more "wise" decisions.

References

1. Wang, X.: Industry 4.0: intelligent industry. Technol. Internet Things (12), 1–3 (2013). (in Chinese)
2. Wahlster, W.: From industry 1.0 to industry 4.0: towards the 4th industrial revolution. Forum Business meets Research (2012). (in Chinese)
3. Yen, C.T., Liu, Y.C., Lin, C.C., et al.: Advanced manufacturing solution to industry 4.0 trend through sensing network and cloud computing technologies. In: 2014 IEEE International Conference on Automation Science and Engineering (CASE), pp. 1150–1152. IEEE (2014). (in Chinese)
4. Peng, W.: The Key Technology and Application of Cloud Computing, pp. 73–75. The People's Posts and Telecommunications Publishing House, Beijing (2010). (in Chinese)
5. Liu, K., Li, A.: Research and implementation of cloud storage based on Hadoop. Micro Comput. Inf. 27(7) (2011). (in Chinese)
6. Dean, J., Ghemawat, S.: MapReduce: simplified data processing on large clusters. Commun. ACM 51(1), 107–113 (2008). (in Chinese)
7. Lin, J., Dyer, C.: Data-intensive text processing with MapReduce. Synth. Lect. Hum. Lang. Technol. 3(1), 1–177 (2010). (in Chinese)
8. Gao, H., Zhai, Y.: Research of mobile learning model based on hadoop. China Audiovisual Education 2011(288). (in Chinese)
9. Borthakur, D.: The hadoop distributed file system: architecture and design. Hadoop Proj. Website 11(2007), 21 (2007). (in Chinese)
10. Cheng, X.: Demands, environment and service of industrial big data under the structure of industrial 4.0. J. Chifeng Uni. (Nat. Sci. Ed.) 2015(4), 14–15. (in Chinese)
11. Shafer, J., Rixner, S., Cox, A.L.: The hadoop distributed filesystem: balancing portability and performance. In: 2010 IEEE International Symposium on Performance Analysis of Systems & Software (ISPASS), pp. 122–133. IEEE (2010). (in Chinese)
12. Shvachko, K., Kuang, H., Radia, S., et al.: The hadoop distributed file system. In: 2010 IEEE 26th Symposium on Mass Storage Systems and Technologies (MSST), pp. 1–10. IEEE (2010). (in Chinese)
13. Lee, J., Kao, H.A., Yang, S.: Service innovation and smart analytics for Industry 4.0 and big data environment. Procedia CIRP 16, 3–8 (2014). (in Chinese)

IoT

A Novel Integrated GSM Balun Design and Simulation

Bing Luo[1], Jianqi Liu[1(✉)], Yongliang Zhang[1], Yanlin Zhang[1], and Bi Zeng[2]

[1] School of Information Engineering, Guangdong Mechanical and Electrical College,
Guangzhou 510515, China
{32999148,936392949,350054049}@qq.com, liujianqi@ieee.org
[2] Guangdong University of Technology, Guangzhou, China
zb9215@gdut.edu.cn

Abstract. Base on the LTCC technology, the high integrated circuit, the excellent performance high frequency characteristic and good reliability can be achieved. This paper depicted theoretical analysis of Balun, and simulated a integrated and small Balun by using HFSS, its center frequency is 900 MHz, band width is 120 MHz, unbalance and balance impedance are 50 Ω, the degree of phase balance is $180 \pm 6°$. The real Balun is manufactured by using microwave ceramics material which its dielectric constant is 8 and loss tangent is 0.0005, By comparing the results of HFSS simulation analysis with the results of experiment, it shows clearly that experimental data is consistent with simulated data. It is discussed that some should be keep watch out in the experiment and simulation, and some technique is presented to control error between experimental data with simulated data.

Keywords: Balun · LTCC · Integrated · HFSS · Simulation and analysis

1 Introduction

With the development of communication technology, the role of radio frequency (RF) device is becoming more and more important. Balun is a research hot spot as one of RF devices. Balun is a three port device, consisted of an unbalanced port and two balanced ports. The signals of the two balanced ports have the same amplitude, but there is a phase shift of 180°. Many circuits need to balance input and output, which is used to reduce the noise and the high harmonics of the circuit, and to improve the dynamic range of the circuit. Balun is widely used in balanced mixer, push-pull amplifier and antenna circuit [1–4].

There is a variety of the form of Balun, generally can be divided into active and passive Balun. Due to the active use of the transistor or other active devices, so it will inevitably produce noise and power consumption. The passive Balun can be divided into three categories, which are the lumped component type, the form of the spiral transformer, and the distribution parameters. The advantages of the lumped element form are small size and light-weight, but not easy to reach 180° of the phase shift and equal output amplitude; The spiral transformer type is only suitable for low frequency and ultra high frequency (UHF), and there is a certain loss; The Balun of distributed parameters is divided into the form of 180° mixed ring and Marchand Balun [5]. In the microwave

© ICST Institute for Computer Sciences, Social Informatics and Telecommunications Engineering 2016
J. Wan et al. (Eds.): Industrial IoT 2016, LNICST 173, pp. 117–124, 2016.
DOI: 10.1007/978-3-319-44350-8_12

band, the 180° mixing loop has a fairly good frequency response, but limit is a large size. It is applied to the radio frequency band from 200 MHz to a few GHz. Due to a better output amplitude and 180° of phase shift, and the bandwidth is wider, so Marchand Balun is used by many designers [6, 7]. However, the Marchand Balun is composed of 1/4 wavelength coupled line, which will occupy a large area, especially in the lower frequency band [8]. In this paper, we design an integrated, compact, light-weight, high reliability, multi-functional and low cost Marchand Balun with the method of multi-layer structure to reduce the bulk.

2 Balun Theory

Figure 1 is a schematic diagram of the Marchand Balun, the electrical length of each transmission line is $\lambda/4$, the signal is inputted/outputted from the unbalanced port and outputted/inputted from the balanced port, the two balanced port phase difference is 180°. The characteristic impedance of the four stage microstrip line and the coupling degree of the coupling section are designed suitably, then, the function of the Balun conversion and impedance transformation can be realized at the same time. Figure 2 depicted schematic diagrams of broadside coupled line Balun section, the coupling section A and B in cross section is completely symmetrical structure, ε_r is the dielectric constant of the dielectric.

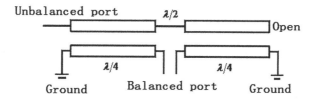

Fig. 1. Marchand microstrip Balun with edge coupling structure

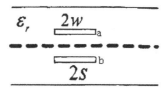

Fig. 2. Marchand microstrip Balun section with edge coupling structure

Marchand Balun equivalent circuit diagram is depicted in Fig. 3. Respectively, k_a, Z_{ac}, k_b, Z_{bc} were the coupling coefficient and characteristic impedance of broadside coupled belt line a and b segment. Z is the output impedance, R is the load. It can be further simplified equivalent circuit, as shown in Fig. 4, supposed $k = k_a = k_b$, Eqs. (1)–(5) can be derived:

$$Z'_1 = Z_{ac}\sqrt{1 - k^2} \tag{1}$$

$$Z'_2 = Z_{bc}\sqrt{1 - k^2} \tag{2}$$

$$Z'_3 = (Z_{ac} + Z_{bc})\frac{k^2}{\sqrt{1 - k^2}} \tag{3}$$

$$Z'_4 = Zk^2 \tag{4}$$

$$R'_5 = Rk^2 \tag{5}$$

from (1)–(5),

$$k = \frac{Z'_3}{\sqrt{Z'_1 + Z'_2 + Z'_3}} \tag{6}$$

$$Z_{ac} = \frac{Z'_1}{\sqrt{1 - k^{2'}}} \tag{7}$$

$$Z_{bc} = \frac{Z'_2}{\sqrt{1 - k^{2'}}} \tag{8}$$

$$Z = \frac{Z'_4}{k^2} \tag{9}$$

$$R = \frac{R'_5}{k^2} \tag{10}$$

And

$$Z_{in} = \frac{Z'^2_1}{-jZ'_2 \cot\theta'_2 + \dfrac{jZ_1Z'_3 \tan\theta'_3}{Z_l + jZ'_3 \tan\theta'_3}} \tag{11}$$

Respectively,

$$Z_l = \frac{Z'^2_4}{R'_5} \tag{12}$$

Obviously, Z'_1, Z'_2, Z'_3, Z'_4, R'_5 were chose appropriately, it can be get the ideal characteristics of broadside coupled microstrip Balun.

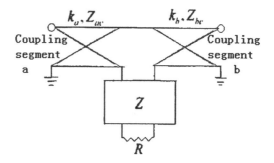

Fig. 3. Balun equivalent circuits 1

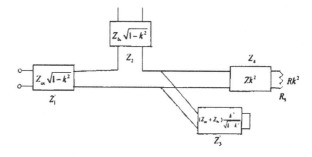

Fig. 4. Balun equivalent circuits 2

3   Analysis of HFSS Simulation

In the actual design process, the calculation process is more complex, so the electromagnetic field simulation software is used for many times to achieve the design goal. HFSS is one of the mainstream FEM 3D high frequency electromagnetic field simulation software. It adopts the advanced technology which is ALPS(Adaptive Lanczos Pade Sweep) and Mode-node conversion, and so on. The FEM method is applied to simulate the electromagnetic field for the passive 3D structure of arbitrary shape. It is applied to the adaptive mesh refinement technique.

Using the HFSS software to simulate, the 3D model structure of Balun is shown with a total of 15 layers in Fig. 5. The first and the second and the third are the ground layer, others is the signal layer.

Application of HFSS parameter optimization of scanning tools, the simulation results of the return loss and two output ports insertion loss are shown in Fig. 6, and the results of phase difference of balance ports are shown in Fig. 7. In 860–960 MHz range, the return loss is more than 13.7 dB, and two port insertion loss is less than 3.9 dB, amplitude imbalance is less than 0.3 dB, the phase difference of balance port is ±4°.

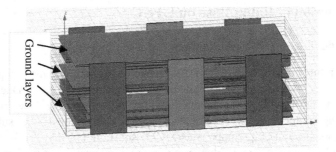

Fig. 5. Balun structure with 15 layers

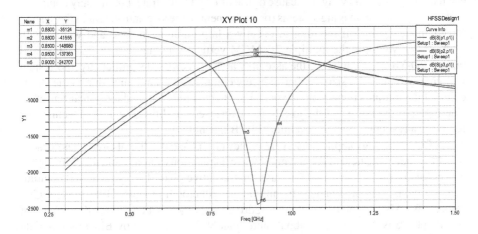

Fig. 6. Return loss and insertion loss of HFSS simulation

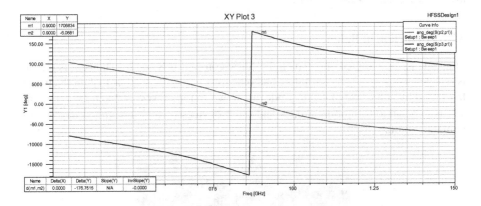

Fig. 7. Phase difference of HFSS simulation

4 Production of Balun Base on LTCC

4.1 Introduction of LTCC Technology

Low temperature co-fired ceramic (LTCC) technology is currently more popular three-dimensional microwave integrated circuit packaging technology [9, 10]. Due to the high number of wiring layers, wiring conductor square resistance, low dielectric constant, low sintering temperature, low thermal expansion coefficient, LTCC substrate has become an ideal multi-chip-components(MCM) substrate [11–13]. At the same time, the resistance, capacitance and inductance can be buried because of the special characteristics of LTCC. LTCC also has excellent high frequency characteristics. The MCM can reduce the size and weight because of use of LTCC substrate, but in many computer systems, high speed imaging systems (or components), the size and weight of the system (or components) is still required to occupy a large number of size and weight with multiple 2D-MCM. If the multiple 2D-MCM vertical stack up to form 3D-MCM and replace 2D-MCM, it can make the assembly area to decreased significantly, the system will greatly reduce the size, further reduce weight. At the same time, due to the shortening of the mutual connection, the parasitic effect is reduced, and the signal transmission is faster, the noise and the loss is also decreased. Additionally, LTCC technology has more advantages than the traditional PCB technology.

LTCC technology applied in low frequency circuit and the digital circuit has decades of history, but in the field of RF and microwave applications abroad from the early 1990s, it is began to research and is widely used in the active phased array radar and communication fields, such as standard mobile phone, Bluetooth module, GPS, PDA, digital camera, WLAN, automotive electronics, drives and other. Among them, the amount of mobile phone is the main part about more than 80 %. Followed by Bluetooth module and WLAN. Due to the high reliability of LTCC products, the application in the automotive electronics is also increasing.

4.2 Production of Balun

Magnetic particle ingredients which dielectric constant is 8 and loss tangent is 0.0005 is selected to casting, screen printing and other processes, the final, actual products was fabricated by sinter and plate electrode, as shown in Fig. 8, size is $2 \times 1.25 \times 0.9$ mm, and is applicable to mobile phones and other devices of high integrated degree and miniaturization.

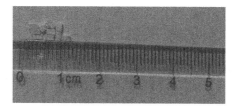

Fig. 8. Balun of implement

5   Results and Discussion

The actual products is tested by using Agilent e5071c network analyzer, and the test results and simulation results are compared and shown in Figs. 9 and 10, solid line is simulation results, and point line is test results, it can be seen that simulation results is consistent with test results from the below picture. In the 820–940 MHz range, the test results show the return loss is more than 12 dB and two port insertion loss are less than 4.1 dB, amplitude imbalance is less than 0.5 dB, and phase unbalance degree is ±6°.

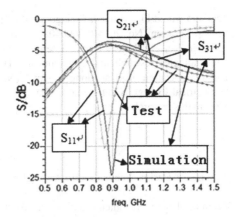

Fig. 9. S parameter comparison of simulation and measurement

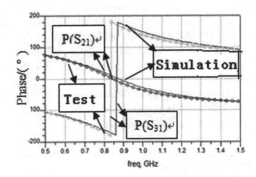

Fig. 10. Phase comparison of simulation and measurement

Obviously, this simulation and design are feasible. The difference between the simulation results and the actual product testing results mainly is the following several reasons: First, LTCC tape casting process of diaphragm have a little thickness error; Second, calibration also exist certain error before the test. Cast film thickness should be measured repeatedly in a real production, and as far as possible to control the film thickness in the admissible error of simulation.

6   Conclusion

This paper designs a kind of Balun with multi-layer structure, its principle is discussed. The results of simulation by using HFSS are compared with tests. The EM simulation and experimental results show a good agreement. The new Balun is small size, light-weight, good balance, low insertion loss, low cost, etc., and can be applied to China Mobile and China Telecom 900 MHz GSM band up-link communication and other systems.

Acknowledgments. The authors would like to thank Guangdong Province Special Project of Industry-University-Institute Cooperation (No. 2014B090904080), 2013 Guangdong Province University High-level Personnel Project (Project Name: Energy-saving building intelligent management system key technologies research and development), Higher Vocational and Technical Education Research Association of Guangdong Province (No. GDGZ14Y033) and Guangdong Mechanical & Electrical College Natural Science Fund (No. YJL2014—042, No. YJKJ2015-2) for their support in this research.

References

1. Sayre, C.W.: Complete Wireless Design. Publishing House of Electronics Industry, Beijing (2010)
2. Pang, J., Jiang, Y.: Design of planar spiral antenna and its wideband Balun. J. Microwaves **s3**, 128–130 (2012)
3. Liu, J., Wan, J., Wang, Q., Deng, P., Zhou, K., Qiao, Y.: A survey on position-based routing for vehicular ad hoc networks. Telecommun. Syst. (2015). doi:10.1007/s11235-015-9979-7
4. Liu, J., Wang, Q., Wan, J., Xiong, J., Zeng, B.: Towards key issues of disaster aid based on wireless body area networks. KSII Trans. Internet Inf. Syst. **7**(5), 1014–1035 (2013)
5. Lin, Q., Zhang, Z., Zhang, B.: Microstrip Balun design. Mod. Radar **24**(10), 61–63 (2004)
6. Jiang, W., Jin, L., Yang, S., Hu, J.: Design of a miniaturized LTCC Balun. Electron. Compon. Mater. **30**(9), 53–56 (2011)
7. Qibo, H., Jianrong, C.: Design of wide band Balun applied to LTCC doubly balanced mixer. Space Electron. Technol. **2**, 107–110 (2010)
8. Marchand, N.: Transmission-line conversion transformers. Electronics **17**(12), 142–145 (1944)
9. Du, Z., Zhu, M., Zhang, H., Guo, T.: Combination of blun and filter based on UWB system. J. Microwaves **2010**(8), 351–354 (2010)
10. Simon, W., Kulke, R., Wien, A., Rittweger, M., Wolff, I., Girard, A., Bertinet, J.-P.: Interconnects and transitions in multilayer LTCC multichip modules for 24 GHz ISM-band applications. In: IEEE MTT-S International Microwave Symposium Digest, vol. 2, pp. 1047–1050 (2000)
11. Yan, W., Fu, P., Hong, W.: Microwave characteristics of bonding interconnects in LTCC microwave MCM. J. Microwaves **19**(3), 30–34 (2003)
12. Yan, W., Yu, S., Fang, X.: Three dimensional integrated microwave modules based on LTCC technology. Acta Electronica Sin. **33**(11), 2009–2012 (2005)
13. Schmuckle, F.J., Jentzsch, A., Heinrich, W., Butz, J., Spinnler, M.: LTCC as MCM substrate: design of strip—line structure and flip-chip interconnects. In: IEEE MTT-S International Microwave Symposium Digest, vol. 2, pp. 1903–1906 (2000)

A Novel Vehicular Integrated Positioning Algorithm

Jianqi Liu[1], Yanlin Zhang[1(✉)], and Bi Zeng[2]

[1] School of Information Engineering,
Guangdong Mechanical & Electrical College, Guangzhou, China
liujianqi@ieee.org, 350054049@qq.com
[2] Guangdong University of Technology, Guangzhou, China
zb9215@gdut.edu.cn

Abstract. The Global navigation satellite system (GNSS) can offer high precise location service for vehicle in open area, but in urban, the satellite signal is obscured by dense building, tunnel. Dead reckoning (DR) system can estimate vehicular position in short period of time, but do not work well as its error accumulation with the passage of time. If the area has been deployed RSU, vehicle can get own position by communication. A single positioning system is not able to offer precise location service in urban, this paper combines with RSU, GNSS and DR and proposes a integrated position system. The integrated system makes use of Federate Kalman Filter (FKF) algorithm to realize information fusion of three systems. The experimental results show the positioning accuracy and anti-jamming capability of RSU/GNSS/DR integrated positioning system is better than a single system.

Keywords: Vehicle positioning · Information fusion · Federate Kalman Filter · Integrated positioning system

1 Introduction

GNSS has some advantages such as high precision, low cost, ease of use, as long as it can receive the satellite signal where the vehicular absolute position can be calculated, and the positioning error will not accumulate with the passage of time. However, in the city, as the satellite signals are obscured by dense buildings, the positioning accuracy decline, when the satellite positioning system is used alone. Even more serious is that the vehicle can not positioning in the tunnel, so the positioning system has lower reliability [1, 2]. Dead reckoning (DR) system is commonly used in the autonomous positioning method for a vehicle. As long as vehicular initial position is set, DR system can calculate the vehicular current position by speed and direction, and can provide high-precision positioning in a short time. If the initial position of the DR system can not be calibrated during a shorter period of time, because of its error accumulation with the passage of time, positioning system's reliability is also declined [3, 4]. With the development of VANET, RSU-based positioning system has been applied [5, 6]. Using three times communication between vehicle and RSU, RSU-based positioning system do not rely on satellites to provide location service for the vehicle, if the positioning

© ICST Institute for Computer Sciences, Social Informatics and Telecommunications Engineering 2016
J. Wan et al. (Eds.): Industrial IoT 2016, LNICST 173, pp. 125–136, 2016.
DOI: 10.1007/978-3-319-44350-8_13

areas have employed RSU. Positioning accuracy is in direct proportion to RSU-employed intensity [7], the more RSU deployment, the higher positioning precision, but the higher the cost. These three systems have advantages and disadvantages, but in a complex environment, a single system can not satisfy the application requirements of current vehicle positioning, using information fusion technology to improve the reliability and accuracy of the positioning system, it will be a viable solution for vehicle location [8].

This paper will combine with RSU, GNSS and DR systems and proposed a integrated positioning system for vehicular location service. In Sect. 2, we introduce the designation of integrated positioning system. Section 3 makes positioning experiment to verify the validity of the proposed algorithm. Section 4 this paper gives a conclusion and discusses some open issues.

2 RSU/GNSS/DR Integrated Positioning Algorithm Designation

In order to solve the vehicle positioning problem in complex urban environment, vehicular positioning solution needs to integrate multiple positioning technologies to realize high-precision, high-reliability positioning. As centralized Kalman Filter integrated algorithm has some problems such as heavy calculation burden, faulted subsystem isolation is difficult, it can not be up to multi-dimensional information fusion vehicular positioning. This section will discuss the RSU/GNSS/DR integrated positioning assisted by FKF algorithm to offer vehicular location service.

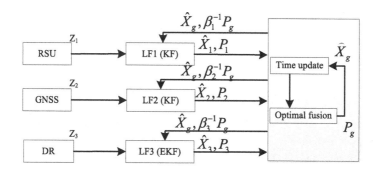

Fig. 1. The framework of RSU/GNSS/DR integrated positioning system

RSU-based positioning subsystem uses a linear Kalman Filter as the Local Filter, names as LF1, the corresponding information sharing coefficient is β_1; GNSS positioning subsystem also uses a linear Kalman filter, names as LF2, the corresponding information sharing coefficient is β_2; DR positioning subsystem uses extended Kalman Filter (EKF), names as LF3, the corresponding information sharing coefficient is β_3; main filter is responsible to information fusion, the framework of RSU/GNSS/DR integrated positioning system is shown in Fig. 1.

2.1   System State Equation and Observation Equation

Using "current" statistic model to describe the statistical distribution of the vehicle acceleration, the significance of this model is that, people merely concern about the "current" probability density of maneuvering acceleration, i.e. current probability of vehicle maneuvering, when a vehicle maneuver in a positive acceleration, which in the next instantaneous acceleration value range is limited, and only in the adjacent range of the current acceleration, which is shown in Eq. (1).

$$a_1(t) = \bar{a}(t) + a(t)$$
$$\dot{a}(t) = -\frac{1}{\tau}a(t) + w(t) \tag{1}$$
$$\dot{a}(t) = -\frac{1}{\tau}a_1(t) + \frac{1}{\tau}\bar{a}(t) + w(t)$$

where, $a_1(t)$ is maneuvering acceleration, its variance is σ^2, $\bar{a}(t)$ is the "current" mean of maneuvering acceleration, $\bar{a}$ is constant in each sampling cycle. $a(t)$ is colored acceleration noise with zero mean, $w(t)$ is White Gaussian Noise with zero mean, τ is correlative time constant of maneuvering acceleration change rate.

The state equation of integrated positioning algorithm is $X = [x_e, v_e, a_e, x_n, v_n, a_n]^T$, where x_e and x_n are the easterly and northerly location component respectively (units: m), v_e and v_n are the easterly and northerly velocity component respectively (units: m/s), a_e and a_n are the easterly and northerly acceleration component respectively (units: m/s^2). The state equation of integrated system can be represented by Eq. (2).

$$\hat{X}(t) = AX(t) + U + W(t) \tag{2}$$

where,

$$A = \begin{bmatrix} 0 & 1 & 0 & 0 & 0 & 0 \\ 0 & 0 & 1 & 0 & 0 & 0 \\ 0 & 0 & -1/\tau_{a_e} & 0 & 0 & 0 \\ 0 & 0 & 0 & 0 & 1 & 0 \\ 0 & 0 & 0 & 0 & 0 & 1 \\ 0 & 0 & 0 & 0 & 0 & -1/\tau_{a_n} \end{bmatrix}, \ U = \begin{bmatrix} 0 \\ 0 \\ \frac{1}{\tau_{a_e}}\bar{a}_e \\ 0 \\ 0 \\ \frac{1}{\tau_{a_n}}\bar{a}_n \end{bmatrix}, \ W(t) = \begin{bmatrix} 0 \\ 0 \\ w_{a_e} \\ 0 \\ 0 \\ w_{a_n} \end{bmatrix}$$

w_{a_e} and w_{a_n} are White Gaussian Noise, whose means are zero, variances are $\sigma_{a_e}^2$ and $\sigma_{a_n}^2$ respectively, τ_{a_e} and τ_{a_n} are easterly and northerly correlative time constant of maneuvering acceleration change rate respectively, $\bar{a}_e$ and $\bar{a}_n$ are easterly and northerly "current" mean of maneuvering acceleration component respectively.

Presupposed the sampling cycle is T, by discretizing continuous system state equation, the discrete system state equation can be derived.

$$X(k) = \Phi(k/k-1)X(k/k-1) + U(k) + W(k) \tag{3}$$

where, $X(k) = [x_e(k), v_e(k), a_e(k), x_n(k), v_n(k), a_n(k)]^T$

$$\Phi(k/k-1) = diag[\Phi_e(k/k-1), \Phi_n(k/k-1)] \tag{4}$$

set $\alpha_e = \frac{1}{\tau_{ae}}$, $\alpha_n = \frac{1}{\tau_{an}}$, $\Phi_e(k/k-1)$ and $\Phi_n(k/k-1)$ are

$$\Phi_e(k/k-1) = \begin{bmatrix} 1 & T & \alpha_e^{-2}(-1+\alpha_e T + e^{-\alpha_e T}) \\ 0 & 1 & (1-e^{-\alpha_e T})\alpha_e^{-1} \\ 0 & 0 & e^{-\alpha_e T} \end{bmatrix},$$

$$\Phi_n(k/k-1) = \begin{bmatrix} 1 & T & \alpha_n^{-2}(-1+\alpha_n T + e^{-\alpha_n T}) \\ 0 & 1 & (1-e^{-\alpha_n T})\alpha_n^{-1} \\ 0 & 0 & e^{-\alpha_n T} \end{bmatrix}$$

$$U(k) = [u_1 \quad u_2 \quad u_3 \quad u_4 \quad u_5 \quad u_6]^T \tag{5}$$

where,

$$u_1 = \left[-T + 0.5\alpha_e T^2 + (1-e^{-\alpha_e T})\alpha_e^{-1}\right]\alpha_e^{-1}\bar{a}_e$$
$$u_2 = \left[T - (1-e^{-\alpha_e T})\alpha_e^{-1}\right]\bar{a}_e$$
$$u_3 = (1-e^{-\alpha_e T})\bar{a}_e$$
$$u_4 = \left[-T + 0.5\alpha_n T^2 + (1-e^{-\alpha_n T})\alpha_n^{-1}\right]\alpha_n^{-1}\bar{a}_n$$
$$u_5 = \left[T - (1-e^{-\alpha_n T})\alpha_n^{-1}\right]\bar{a}_n$$
$$u_6 = (1-e^{-\alpha_n T})\bar{a}_n$$

2.1.1 RSU Observation Equation

According to positioning theory of the RSU-based subsystem, we set the system state variable $X_1 = X$, subsystem state equation is same as system state equation.

The observation values, who is output of RSU-based positioning subsystem, are easterly and northerly coordinates component e_{obs} and n_{obs} (units: m), the observation equation is represented by Eq. (6)

$$Z_1(k) = H_1(k)X_1(k) + V_1(k) \tag{6}$$

where,

$$Z_1(k) = \begin{bmatrix} e_{obs}(k) \\ n_{obs}(k) \end{bmatrix}, \quad V_1(k) = \begin{bmatrix} w_e(k) \\ w_n(k) \end{bmatrix}, \quad H_1(k) = \begin{bmatrix} 1 & 0 & 0 & 0 & 0 & 0 & 0 & 0 \\ 0 & 0 & 0 & 1 & 0 & 0 & 0 & 0 \end{bmatrix},$$

$w_e(k)$ and $w_n(k)$ are White Gaussian Noise, whose means are zero, variance are σ_e^2 and σ_n^2 respectively, $w_e(k)$ and $w_n(k)$ refer to easterly and northerly position measurement noise measured by the RSU positioning device, measurement noise covariance matrix is represented by $R(k)$, where, $R(k) = diag(\sigma_{a_e}^2, \sigma_{a_n}^2)$.

2.1.2 GNSS Observation Equation

GNSS subsystem is similar to RSU-based subsystem, we set the system state variable $X_2 = X$, subsystem state equation is same as system state equation. The observation values are easterly and northerly coordinates component e_{obs} and n_{obs}, the observation equation can be represented by Eq. (7).

$$Z_2(k) = H_2(k)X_2(k) + V_2(k) \tag{7}$$

where,

$$Z_2(k) = \begin{bmatrix} e_{obs}(k) \\ n_{obs}(k) \end{bmatrix}, \quad V_2(k) = \begin{bmatrix} w_e(k) \\ w_n(k) \end{bmatrix}, \quad H_2(k) = \begin{bmatrix} 1 & 0 & 0 & 0 & 0 & 0 & 0 & 0 \\ 0 & 0 & 0 & 1 & 0 & 0 & 0 & 0 \end{bmatrix}$$

$w_e(k)$ and $w_n(k)$ are easterly and northerly measurement error, whose means are zero, variance are σ_e^2 and σ_n^2 respectively. Measurement noise covariance matrix is $R(k) = diag(\sigma_{a_e}^2, \sigma_{a_n}^2)$.

2.1.3 DR Observation Equation

The state variable of DR subsystem is $X_3 = X$, the observation values, who are the output of angular rate gyroscope ω and the output of the vehicle odometer S in a sampling cycle, the subsystem observation equation can be represented by Eq. (8).

$$Z_3(t) = h_3(t, X_3(t)) + V_3(t) \tag{8}$$

where,

$$Z_3 = \begin{bmatrix} \omega \\ s \end{bmatrix} = \begin{bmatrix} \frac{\partial}{\partial t}\left[\tan^{-1}\left(\frac{v_e}{v_n} \right) \right] \\ \varphi T \sqrt{v_e^2 + v_n^2} \end{bmatrix} + \begin{bmatrix} \varepsilon_\omega \\ \varepsilon_s \end{bmatrix}$$

after reduction,

$$Z_3 = \begin{bmatrix} \omega \\ s \end{bmatrix} = \begin{bmatrix} \frac{v_n a_e - v_e a_n}{v_e^2 + v_n^2} \\ T \sqrt{v_e^2 + v_n^2} \end{bmatrix} + \begin{bmatrix} \varepsilon_\omega \\ \varepsilon_s \end{bmatrix}$$

A calibration factor φ is assumed as 1, ε_ω is the gyroscope drift, approximate to $(0, \sigma_\omega^2)$ White Gaussian Noise, ε_s is the observation noise of odometer output S, approximate to $(0, \sigma_s^2)$ White Gaussian Noise, $R(k)$ is measurement noise covariance matrix, $R(k) = diag(\sigma_{a_\omega}^2, \sigma_{a_s}^2)$. After discretizing continuous observation equation, the discrete observation equation is shown in Eq. (9)

$$Z_3(k) = h_3[k, X_3(k)] + V_3(k) \tag{9}$$

DR employs EKF as local filter, and lets $h_3[X_3(k)]$ to expand in the vicinity of $\hat{X}(k/k-1)$ according to Taylor series. Ignoring secondary or higher order entry, DR can obtain linear observation equation.

$$Z_3(k) = H_3(k)X_3(k) + h_3\left[k, \hat{X}_3(k/k-1)\right] - H_3(k)\hat{X}_3(k/k-1) + V_3(k) \qquad (10)$$

where,

$$H_3(k) = \left.\frac{\partial h_3[X_3(k)]}{\partial X_3(k)}\right|_{X_k=\hat{X}_{k,k-1}} = \begin{bmatrix} 0 & h_1 & h & 0 & h_3 & h_4 \\ 0 & h_5 & 0 & 0 & h_6 & 0 \end{bmatrix}$$

$$h_1 = \frac{\hat{a}_n(k/k-1)\hat{v}_e^2(k/k-1) - 2\hat{v}_e(k/k-1)\hat{v}_n(k/k-1)\hat{a}_e(k/k-1) - \hat{a}_n(k/k-1)v_n^2(k/k-1)}{\left[v_e^2(k/k-1) + v_n^2(k/k-1)\right]^2}$$

$$h_2 = \frac{\hat{v}_n(k/k-1)}{v_e^2(k/k-1) + v_n^2(k/k-1)}$$

$$h_3 = \frac{\hat{a}_e(k/k-1)\hat{v}_e^2(k/k-1) - 2\hat{v}_e(k/k-1)\hat{v}_n(k/k-1)\hat{a}_e(k/k-1) - \hat{a}_n(k/k-1)v_n^2(k/k-1)}{\left[v_e^2(k/k-1) + v_n^2(k/k-1)\right]^2}$$

$$h_4 = \frac{\hat{v}_e(k/k-1)}{v_e^2(k/k-1) + v_n^2(k/k-1)}$$

$$h_5 = \frac{T\hat{v}_e(k/k-1)}{\sqrt{v_e^2(k/k-1) + v_n^2(k/k-1)}}$$

$$h_6 = \frac{T\hat{v}_n(k/k-1)}{\sqrt{v_e^2(k/k-1) + v_n^2(k/k-1)}}$$

2.2   Time Update and Measurement Update of Subsystem

LF1 and LF2 subsystem time update and measurement update equation:

$$\hat{X}_i(k/k-1) = \Phi(k/k-1)\hat{X}_i(k-1) + U(k-1) \qquad (11)$$

$$P_i(k/k+1) = \Phi(k/k-1)P_i(k-1)\Phi^T(k/k-1) + Q(k-1) \qquad (12)$$

$$K_i(k) = P_i(k/k-1)H_i^T(k)\left[H_i(k)P_i(k/k-1)H_i^T(k) + R_i(k)\right]^{-1} \qquad (13)$$

$$\hat{X}_i(k) = \hat{X}_i(k/k-1) + K_i(k)[Z_i(k) - H_i(k)X_i(k/k-1)] \qquad (14)$$

$$P_i(k) = [1 - K_i(k)H_i(k)]P_i(k/k-1) \qquad (15)$$

State transition matrix: the prediction value of acceleration is considered as the mean of "current" acceleration, i.e.

$$\bar{a}_e(k) = \hat{a}_e(k/k-1), \quad \bar{a}_n(k) = \hat{a}_n(k/k-1)$$

Equation (11) can be simplified to Eq. (16)

$$\hat{X}_1(k/k-1) = \Phi(k/k-1)\hat{X}_1(k-1) \tag{16}$$

where, $\Phi(k/k-1) = diag[\Phi_e(T), \Phi_n(T)]$, i.e.

$$\Phi_e(T) = \begin{bmatrix} 1 & T & T^2/2 \\ 0 & 1 & T \\ 0 & 0 & 1 \end{bmatrix}, \quad \Phi_n(T) = \begin{bmatrix} 1 & T & T^2/2 \\ 0 & 1 & T \\ 0 & 0 & 1 \end{bmatrix}$$

System noise covariance matrix: $Q(k-1)$ is the discrete matrix of system noise covariance matrix Q

$$Q(k-1) = \begin{bmatrix} \frac{2\sigma_{a_e}^2}{\tau_{a_e}} Q_e(T) & 0_{3 \times 3} \\ 0_{3 \times 3} & \frac{2\sigma_{a_n}^2}{\tau_{a_n}} Q_n(T) \end{bmatrix}$$

where,

$$Q_e(T) = Q_n(T) \approx \begin{bmatrix} T^5/20 & T^4/8 & T^3/6 \\ T^4/8 & T^3/3 & T^2/2 \\ T^3/6 & T^2/2 & T \end{bmatrix}.$$

DR subsystem employs EKF as the local filter LF3, the computing process is different from the LF1 and LF2, which is shown in Eq. (17).

$$\begin{aligned}
&\hat{X}_3(k/k-1) = \Phi(k/k-1)\hat{X}_3(k-1) \\
&P_3(k/k+1) = \Phi(k/k-1)P_3(k-1)\Phi^T(k/k-1) + Q(k-1) \\
&K_3(k) = P_3(k/k-1)H_3^T(k)\left[H_3(k)P_3(k/k-1)H_3^T(k) + R_3(k)\right]^{-1} \\
&\hat{X}_3(k) = \hat{X}_3(k/k-1) + K_3(k)[Z_3(k) - h_3(k, X_3(k/k-1))] \\
&P_3(k) = [1 - K_3(k)H_3(k)]P_3(k/k-1)
\end{aligned} \tag{17}$$

where, $Q(k-1)$ and $\Phi(k/k-1)$ is same as one of LF1 and LF2.

2.3   Global Information Integration and Information Sharing

Global information integration:

$$\begin{aligned}
&\hat{X}_g(k) = P_g(k)\left[P_1^{-1}(k)\hat{X}_1(k) + P_2^{-1}(k)\hat{X}_2(k) + P_3^{-1}(k)\hat{X}_3(k)\right] \\
&P_g^{-1}(k) = P_1^{-1}(k) + P_2^{-1}(k) + P_3^{-1}(k) \\
&Q_g^{-1}(k) = Q_1^{-1}(k) + Q_2^{-1}(k) + Q_3^{-1}(k)
\end{aligned} \tag{18}$$

Information sharing:

$$\begin{cases} Q_i(k) = \beta_i^{-1} Q_g(k) \\ P_i(k) = \beta_i^{-1} P_g(k) & i = 1, 2, 3 \\ \hat{X}_i(k) = \hat{X}_g(k) \end{cases} \tag{19}$$

where, $\beta_1 + \beta_2 + \beta_3 = 1$.

3   Testing and Analysis

3.1   The Vehicular Testing Scenario

The testing scenario is shown in Fig. 2, the route from the start point to the end point along the arrow direction, khaki route is the predetermined moving route, the whole testing route can be divided into three zones. First, open area, in where the satellite signal is not blocked; second, tunnels, in where almost all of the satellite signal is blocked; and third, the are is beside buildings, in where part of the satellite signal is blocked. This testing scenario simulates different situation of the vehicle on the road, the reliability and positioning accuracy of integrated positioning system can be test.

We mainly simulate the RSU-failed situation in first area, so we do not deploy RSU in first area, the vehicle depends on multimode GNSS subsystem to offer location service. In second area, we test the positioning performance of integrated system in GNSS-failed situation, so we deploy 7 RSUs (red point in Fig. 2), and preset the UTM

Fig. 2. The testing scenario (Color figure online)

coordinates. In third area, we test the positioning performance of GNSS/DR in RSU-failed situation, which will be influenced by building.

3.2 Vehicular Positioning and Result Analysis

3.2.1 Testing Process

According to predetermined route in testing scenario, the whole test is divided into GNSS-only positioning experiment, RSU-only positioning experiment, RSU/GNSS/DR integrated positioning system experiment. The test process is as follows:

1. **Device Initialization.** Set the UART parameter (the port number: COM6 and baud rate: 4800Bd) for multimode GNSS receiver, set the GNSS data storage directory, set the DSRC receiver channel (CH182) and data storage directory of vehicular RSU-based positioning subsystem, set the initial position of the DR subsystem.
2. **Time Calibration.** In order to maintain time synchronization in GNSS and DSRC, calibration of both time reference.
3. **Location data acquisition.** Vehicle tracks the khaki line at a steady speed of 20 km/h, get location data by integrated positioning system.
4. **Coordinate transformation.** GPS output format is latitude and longitude coordinates, location data format of RSU-based subsystem is UTM coordinate. Since the coordinate system is different in subsystems, we need to achieve coordinate transformation in order to calculate the vehicular coordinates.
5. **Information fusion.** Using output of three positioning subsystems, the integrate system calculates vehicular coordinates by information fusion algorithm.
6. **Trajectory display.** Using the output data of integrated system to generate KML files, and then displays the testing trajectory on Google Earth software.

3.2.2 Positioning Result and Analysis

From the Fig. 3, we can know that, in fist area, the moving trajectory of the integrated system is much same as GNSS' one. The multimode GNSS receiver can receive both BeiDou System (BDS) and Global Positioning system (GPS) satellite signal, and can capture 8–9 satellites simultaneously during the test process. The RSU-based subsystem is considered as a fault system and isolated by integrated system. The positioning output of GNSS/DR play a leading role, the 2DRMS of positioning accuracy is not larger than 2 m.

In second area, the situation is just the opposite. The multimode GNSS receiver can not receive the signal in tunnel. From the green trajectory, we can see that the positioning accuracy decline rapidly. The GNSS is regarded as a fault subsystem, and is isolated by integrated system. The positioning output of RSU/DR play a leading role, and the accuracy is better than the accuracy of RSU-only positioning system, the 2DRMS of positioning accuracy is not larger than 2 m.

In third area, as the RSU communication range do not cover this area, so RSU subsystem is isolated by integrated system, the GNSS/DR offer the positioning service. But the performance of GNSS/DR is influenced by building, the positioning output deviate the road. At the front half of the trip in the third area, the positioning error is corrected by DR subsystem. But at the latter half of the trip, the error is accumulated gradually as time goes on, the DR subsystem is regarded as fault system by fault detected algorithm, GNSS subsystem does well only in this situation. If we deploy RSUs for vehicular positioning in this special area, the accuracy will be improved (Fig. 4).

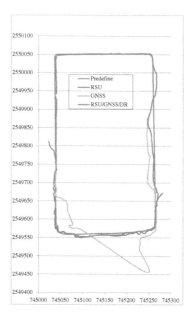

Fig. 3. The comparison chart of positioning accuracy (Color figure online)

Fig. 4. The result of RSU/GNSS/DR integrated positioning

From the positioning testing and analysis abovementioned, the integrated positioning system is up to vehicular positioning in urban, and the positioning accuracy can meet road-level requirement.

4 Conclusion and Discussion

Federal Kalman Filter has been identified as the common mode filter of next-generation navigation system by Air Force, due to its flexible design, a small amount of calculation, good fault tolerance. FKF, who is composed of several sub filters and a main filter, is a two-step cascade of decentralized filtering method, which uses information sharing to achieve global optimization. For vehicle navigation and positioning, FKF has the following advantages over centralized filtering method.

Firstly, the filtered fault tolerance is better. Once a subsystem malfunction, FKF can easily detect and isolate faulted subsystem, and can employ the remaining subsystem to reconstruct the new filter, and calculate the required filter solution.

Secondly, the synthesis filtering algorithm is simper, has lighter computation burden, and less data communication, which is conductive to real-time execution in onboard devices. As a result of using the decentralized two-stage filter program, so the dimension of the filter has been greatly decreased, and reduce the amount of calculation. Vehicular embedded device computing power is weaker than personal computer, the Federal filter can run on the resource-constrained hardware platform.

Thirdly, good scalability. We can be easy to extend this algorithm by adding a map matching algorithm or vision-based positioning algorithm in future study.

Acknowledgments. The authors would like to thank Guangdong Province Special Project of Industry-University-Institute Cooperation (No. 2014B090904080), 2013 Guangdong Province University High-level Personnel Project (Project Name: Energy-saving building intelligent management system key technologies research and development) and the Project of Guangdong Mechanical & Electrical College (No. YJKJ2015-2) for their support in this research.

References

1. Ben-Moshe, B., Elkin, E., Levi, H., Weissman. A.: Improving accuracy of GNSS devices in urban canyons. In: 23rd Canadian Conference on Computational Geometry, Toronto, pp. 399–405 (2011)
2. Hein, G.W.: From GPS and GLONASS via EGNOS to Galileo-positioning and navigation in the third millennium. GPS Solutions **3**, 39–47 (2000)
3. Hide, C., Moore, T., Smith, M.: Adaptive Kalman filtering for low-cost INS/GPS. J. Navig. **56**, 143–152 (2003)
4. Ali, J., Jiancheng, F.: SINS/ANS/GPS integration using federated Kalman filter based on optimized information-sharing coefficients. In: AIAA Guidance, Navigation, and Control Conference and Exhibit, San Francisco pp. 1–13 (2005)
5. Liu, J., Wan, J., Wang, Q., Deng, P., Zhou, K., Qiao, Y.: A survey on position-based routing for vehicular ad hoc networks. Telecommun. Syst., 1–16 (2015)

6. Liu, J., Wan, J., Wang, Q., Li, D., Qiao, Y., Cai, H.: A novel energy-saving one-sided synchronous two-way ranging algorithm for vehicular positioning. Mob. Netw. Appl. **20**, 661–672 (2015)
7. Liu, J., Wan, J., Wang, Q., Zeng, B., Fang, S.: A time-recordable cross-layer communication protocol for the positioning of vehicular cyber-physical systems. Future Gener. Comput. Syst. **56**, 438–448 (2015)
8. He, J., Yuan, X.-L., Zeng, Q., Liu, W.-K.: Study on GPS/BDS/GLONASS combined single point positioning. Sci. Surv. Mapp. **39**, 124–128 (2014)

Electronic Commerce Platform of Manufacturing Industry Under Industrial Internet of Things

Yingying Wang[1], Hehua Yan[1(✉)], and Jiafu Wan[2]

[1] Guangdong Mechanical and Electrical College, Guangzhou, China
wyybaby@163.com, hehua_yan@126.com
[2] South China University of Technology, Guangzhou, China

Abstract. With the development and evolution of industrial information technology, the market competition intensifies increasingly of traditional manufacturing, the polarization speed up. So enterprises must change from the original traditional way to the Internet and innovate business models. In this article, we first briefly describe the relationship between the industrial Internet of things and electronic commerce (E-Commerce), and the current situation and development trend of E-Commerce in the manufacturing industry. Next the new model of the manufacturing E-Commerce platform is explained. Then we give a new framework of the E-Commerce platform. We hope to inspire more technological development and progress for Manufacturing E-Commerce.

Keywords: Industrial Internet of things · E-Commerce platform · New model · C2M · B2B

1 Introduction

The entire world is in the transition from the old to the new. The traditional business structure is being disintegrated little by little. The first wave of impact strength is the Internet and E-commerce, the second wave is "Industry 4.0". "Industry 4.0" depicts a highly flexible and intelligent manufacturing model with real-time, effective communications between people, equipment and products, and the core is Industrial IoT [1, 2]. What relationship is between "E-Commerce" and "Industrial IoT"? As the current situation shows, E-Commerce solves the better consumption, and Industrial IoT is in order to better production. "Production" and "consumption" are the back and front of the enterprise respectively. That is to say, both transform the enterprise from the back and front. But must emphasize this point: in the future production and consumption is unified, the starting point is different, but the destination is the same. To reshape the core competitiveness, manufacturing enterprises need to combine the Internet marketing and Internet IoT together, to pull production and supply based on customization, and to connect the materials, smart factories, logistics and customers into a whole.

© ICST Institute for Computer Sciences, Social Informatics and Telecommunications Engineering 2016
J. Wan et al. (Eds.): Industrial IoT 2016, LNICST 173, pp. 137–143, 2016.
DOI: 10.1007/978-3-319-44350-8_14

2 Current Situation and Developing Trend of Manufacturing E-Commerce

In recent years, there has been a lot of manufacturing e-commerce platform. There are three main models: the first E-Commerce platform is constructed by the third party service provider, the second is the purchase platform of the industry vertical portal, and the third is the online mall of the manufacturers [3, 9]. Although the existing business platform achieve online shopping, but does not really promote large-scale development of industry e-commerce. Within the next few years, a large number of self-operated intelligent production and service system will appear, and to promote the transformation and upgrading of manufacturing industry towards intelligentization, servitization, datamation and networking. At the same time, Industrial IoT will also reshape the existing industrial value chain system, "No channel" will be the trend of industrial products sales, which will bring enormous impact to the traditional product vendors, but also give the manufacturing e-commerce a huge business opportunities.

3 Key Points of the New Model for Manufacturing E-Commerce Platform

In view of the current situation and existing problems of manufacturing E-Commerce, and the analysis of the demand for enterprise procurement, combining the present industrial development background, there are six main points to be explored.

(1) Supply chain collaboration: due to the industry products complexity, large price fluctuations, delivery instability and the need to eliminate the influence of the traditional marketing channel mode, so it is essential to the success of the platform to integrate the resources of the upper end of the industry chain and achieve business collaboration. Industry associations or third party service provider or large-scale manufacturing enterprises lead several manufacturers corporately develop the new business service model, in order to eliminate the crisis of confidence of the end users, and ensure the accuracy of the information, and supply quality, delivery timeliness and price stability. Moreover, to achieve rapid business collaboration among the suppliers, manufacturers, platform and customers, to help enterprises quickly realize the full range of management and monitoring for business flow, information flow, capital flow and logistics by the platform [4]. At the same time, the platform support information dissemination, supplier record, bid quotes, online assessment, order generation, quality inspection, and contract payment processes etc., to achieve an open, fair and reasonable procurement. The basic structure of supply chain is shown in Fig. 1.

(2) Full life cycle management: With the rapid development of Internet, information technology gradually affects all aspects of manufacturing and product life cycle. Product life cycle management refers to the information and process management of product from the requirement, planning, design, production, distribution, use, operation, maintenance, until the recycling and disposal of the entire life cycle. It is a technology, but also a manufacturing concept [5]. Traditional E-Commerce platform can only provide shopping guide, order, payment and logistics service, due to the manufacturing industry

Fig. 1. A basic structure of supply clain

product technical complexity and the application relevance, the current service mode can't meet the demand of users, and the new manufacturing E-Commerce platform must be able to provide the full life cycle management and service and provide online and offline support with taking the customer demand as the center, which requires the platform to gather more manufacturing resource to participate in the formation of the collaboration chain covering the full life cycle, to fully mobilize the enthusiasm and creativity of all kinds of subjects, and to meet the diverse needs of users with rapid response.

(3) Sub themes scene service: Because enterprise objects are numerous, enterprise procurement requirements are different and technical services are with different emphasis, it is necessary to set up special area, to provide the corresponding service, and to carry out personalized marketing and promotion according to the needs of different enterprise groups. At the same time, the platform should take industry applications as the center, to establish the full range service of scene navigation and the virtual application experience [6]. Scene navigation may include equipment for technological transformation, new construction, debugging technology, spare parts procurement. Different scenarios are composed by different solutions including program description, system architecture, technical characteristics and device list etc., and partial solutions provide users with the best shopping experience by virtual scene. Users can clearly see the composition of the solution, dynamic effect, and the hardware and software configuration list, but also can quickly inquire about product details, recent sales and alternative products to quickly enter the purchase link.

(4) Professional system support: With the coming of the industrial 4.0, intelligent manufacturing and service system will be produced in great quantities, the future manufacturing E-Commerce platform need to dock with the professional system, such as equipment monitoring system, maintenance system, mobile inspection system etc. Oriented application system of smart factory (such as APP) will monitor the real-time operation of equipment and the fault diagnosis, generating device replacement, installation, debugging and other technical requirements, so that the E-Commerce platform quickly make anticipation judgment and initiative service [7].

(5) Combined with Internet financial: According to the current problems for enterprise users such as a large amount of procurement, complex process links, slow payment, etc. In order not to affect the capital chain operations, we can use the new financial model with Internet based on the purchaser, the supplier and the platform protocol, to flexibly adopt the current Internet financial means, including crowd funding project, order financing, credit payment, secured transaction etc. The platform can also integrate the fourth party arbitration, insurance, bank and other resources to do a good job in risk control.

(6) Personalized customization: Starting from the overall benefit of the supply chain, based on the customer's perspective, the platform meet personalized customer demand and achieve a single product of mass customization and product family for mass customization. On the platform, customers can communicate with the manufacturer directly and propose customization demand. Manufacturers can achieve zero cost financing, zero cost promotion and zero inventory and easily obtain mass production orders by the pre-sale mode of the platform, then trigger the procurement, design, production and delivery of a series of processes. At the same time, it can become a reality to structure the trans-regional dynamic enterprises union, not only can help the enterprises to share design and manufacturing resource and optimize configuration, but also help to improve the rapid reaction and competitive ability. Moreover, the enterprises will be able to cooperate with the leading enterprises and produce fully their own advantages [8].

4   Architecture Design of Manufacturing E-Commerce Platform

Driven by industrial development, based on the current manufacturing industry development model, we explore the design of service-oriented architecture for manufacturing E-commerce platform. The architecture (in Fig. 2) may be composed of five parts, include user access layer, information portal layer, application service layer, management layer, technical support layer.

(1) User access layer: In support of multi-mode access, provide personalized mobile application system (APP) for the users, and accelerate the integration of mobile Internet technology. Establish the service space for the enterprises, which not only include the common functions such as product catalog, product selection, order processing, secure payment, logistics inquiries, delivery querying and after-sales service, but also include

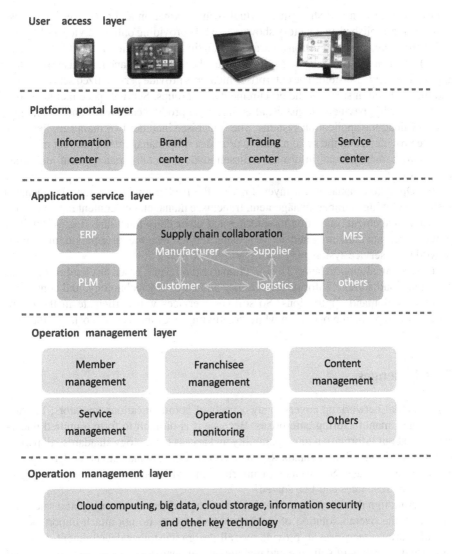

Fig. 2. A architecture of manufacturing E-commerce platform

the value-added services, for examples, equipment inspection, social network, information dissemination, bidding and purchasing, group purchasing. The use can personalize information and functions, and integrate professional manufacturing service system.

(2) Platform portal layer: The platform portals should strengthen fundamentally scene navigation and application experience, break the traditional E-Commerce model and establish information center, brand center, trading center and service center. We should focus on the unique characteristics of the industry to create a professional services center, establish the solutions thematic area according to the direction of technology and

industry application and strengthen virtual scene navigation and experience services. Moreover, establish virtual factory showcase, while providing online service tools, such as 3D browsing, online selection contrast, association query, authentic identification etc.

(3) Application service layer: Mainly for member enterprises and franchisee provides application service system. The enterprises members provide personalized service space and the application store for the procurement user groups. Service space include shopping cart, order, payment, logistics, after-sales and product catalogues and so on, also can download the application systems. The franchisee includes two main suppliers and service providers, and the system mainly provides shop management, product management, order management, information dissemination, account management and other functions.

(4) Operation management layer: Provide the background operation management system, including member management, franchisee management, content management, service management, operation monitoring and other daily functions. It is essential to provide accurate marketing and active sensing service with the information integration of workshop service system.

(5) Technical support layer: The E-Commerce platform is based on cloud computing, big data, cloud storage, information security and other key technology, and constantly develop on-demand applications. So software services will participate in the entire process of e-commerce operation while greatly improves the user viscosity of the web site.

5 Challenges

(1) Industrial networking covers many aspects of communications operators, Internet companies, manufacturing enterprises. Because it is difficult to form a unified understanding about information interoperability and access technology standards, all parties lack the full integration of standards, business processes, management models, knowledge and experience. So "islands of information" will come into being, resulting in information resources can't be shared.

(2) Although the services provide Information products development and sales, but most lack the overall solution of industry segmentation, do not attach importance to manufacturing enterprises to provide overall design, customer development, configuration and operation of software and maintenance management service as a whole, and are located in the design, manufacture and management process. The advanced services of manufacturing value chain are obviously insufficient.

(3) Supply chain collaboration is not only a point to point information exchange and sharing. To achieve information sharing among all members of the supply chain, in response to customer demand, the flexible and collaborative supply chain is a must. The information platform based on the Internet of things can collect and perceive all kinds of information, analyze and process the information, and form a variety of collaborative decision-making.

(4) At present, in the global range data and user information leakage of network security issues have become increasingly prominent, but also a threat to the

manufacturing industry of information security. With the manufacturing industry increasingly using mobile devices and mobile APP, although the efficiency of collaborative work has been improved, but the data security of public WIFI and mobile networks will become particularly prominent, lost mobile device will not only lead to private disclosure of information, so that more enterprises have data leakage risks.

6 Conclusions

In this paper, combined with the development trend of "Industry 4.0", explore a kind of service mode, which is based on supply chain collaboration, full life cycle management, personalized customization and other key points, in order to solve the current industry development bottleneck. At the same time, gives the reference model of the E-Commerce platform, and comprehensively enhance the business collaboration between users, platforms and manufacturers, hoping to promote the specialization and scale development of the E-Commerce platform.

Acknowledgment. This work is partially supported by the National Natural Science Foundations of China (Nos. 61572220 and 61262013), the Fundamental Research Funds for the Central Universities (No. 2015ZZ079), and the Natural Science Foundation of Guangdong Province, China (No. 2015A030313746).

References

1. Wang, S., Wan, J., Zhang, D., Li, D., Zhang, C.: Towards the smart factory for industrie 4.0: a self-organized multi-agent system assisted with big data based feedback and coordination. Elsevier Comput. Netw. (2016). doi:10.1016/j.comnet.2015.12.017
2. Wang, S., Wan, J., Li, D., Zhang, C.: Implementing smart factory of industrie 4.0: an outlook. Int. J. Distrib. Sens. Netw. **2016**, 10 (2016). doi:10.1155/2016/3159805
3. Hippen, B.: Research on the E-commerce model of small and medium enterprises. Appl. Mech. Mater. **20**, 687–691 (2014)
4. Wang, P., Ge, S., Wang, N., Ren, N.: Construction method of collaborative value chain in cloud computing environment. J. Jiangsu Univ. Sci. Technol. (Nat. Sci. Ed.) **29**(6), 585–590 (2015)
5. Shen, J., Zhou, R.: Research on PLM system framework and key technologies. J. Nanjing Univ. Aeronaut. Astronaut. **35**(5), 565–571 (2003)
6. Chen, M., Xu, C., Yu, L.: Design and implementation of intelligent navigation system based on 3D website. Comput. Eng. Des. **31**(20), 4438–4441 (2010)
7. Li, T.: Mobile smart terminal technology industry development elements. Inf. Commun. Technol. **6**(4), 7–11 (2012)
8. Liu, Q., Wan, J., Zhou, K.: Cloud manufacturing service system for industrial-cluster-oriented application. J. Internet Technol. **15**(3), 373–380 (2014)
9. Li, G.: The application research of electronic commerce for industrial design industry. Ind. Sci. Tribune **6**, 100–101 (2012)

A Secure Privacy Data Transmission Method for Medical Internet of Things

Heping Ye, Jie Yang, Junru Zhu, Ziyang Zhang, Yakun Huang,
and Fulong Chen[✉]

Department of Computer Science and Technology, Anhui Normal University,
189 Jiuhua South Road, Wuhu 241002, Anhui Province, People's Republic of China
long005@mail.ahnu.edu.cn

Abstract. With the improvement of people's living level and the rapid development of information, people put forward higher requirements for medical standard. Effective combination between traditional medical system and modern communication technologies promotes the medical level more intelligent. The medical system involves a large number of data, which contains all kinds of information. Therefore, the patient's information is facing the risk of data leakage and privacy information destruction in the transmission process. In order to effectively protect the patient's privacy information, this paper presents a secure data transmission method for privacy data of Medical Internet of Things in three aspects: the transmission model of medical data, the registration authentication and key agreement between the Gateway-node and the Server, and Multi-path transmission mechanism. The theoretical analysis shows that the transmission model could effectively ensure the security of the patient's privacy information.

Keywords: Medical Internet of Things · Privacy data · Transmission method

1 Introduction

The rapid economic development has led to the deterioration of the natural environment upon which the survival of people's health under unprecedented threat. Various non-predictability of diseases have sprung up on the patients so that the patient's illness makes it painful bring the demand for medical services growing. However limited traditional medical service resources and uncertainty treatment time urge people to begin to look for better health service to make up for the lacking of available resources. In [1], a cardiac function in real-time monitoring system that can measure heart rate and other vital signs data, then serving data to the medical center for treatment via Bluetooth communications or wireless networking technologies. Zhang mentions that obtaining data by remote sleeping monitoring could effectively help doctors diagnose disease, and adjust the pillow without affecting the premise of sleep to let patient get the timely healthcare [2].

© ICST Institute for Computer Sciences, Social Informatics and Telecommunications Engineering 2016
J. Wan et al. (Eds.): Industrial IoT 2016, LNICST 173, pp. 144–154, 2016.
DOI: 10.1007/978-3-319-44350-8_15

Mni proposes that today's medical development should take a new information technology way while cleverly epitomizing the medical things meaning. He points out the key technologies in the medical field, analyzes and presents various models about medical data from generation to storage [3].

Due to the huge amount of medical data, extensive medical data sources, and various identification information which involve user privacy, once medical data loses or tampers, leakage will occur. [4,5] have presented that tags will be scanned while users are not aware of what readers do, it will easily result in the destruction of personal privacy, and it will cause the items of information suffering from attacking between Local Servers and Remote Servers. Therefore, we propose a secure privacy data transmission method for Medical Internet of Things.

2   Related Works

Facing with the large number of heterogeneous data in Medical Internet of Things, the problem is how to ensure the security of such data in the remote transmission. It has been always the focus of academic research. Ning [6] considers a variety of secure factors of Internet of Things, and composes that compromise must be existed between privacy strength and specific business needs. Namely, it needs us to custom privacy policy moderately on the basis of business needs as much as possible to protect users' privacy.

Wu introduces that the data protection methods [7] using lightweight cryptographic algorithms in most Internet of Things applications. Du [8] proposes a probabilistic key sharing scheme suitable for WSNS to share. The same communication key exists between any two nodes is p and security is not guaranteed. Song studies the secure and reliable transmission scheme SPS based on Internet of Things [9]. He presents a cooperative transmission mechanism and the rate selection algorithm based on the channel state in order to transmit data effectively and reliably. [10] Lamport first proposes safe way Hash function Encryption users.

Kothmayr proposes an end-and-end mutual authentication mechanism of Internet of Things based on DTLS protocol. The mechanism is based on the existing Public-key encryption algorithm, which is vulnerable to suffer from middle attack because of no three session process [11]. Groce [12] introduces a provably secure PAKE protocol standard model, but there is no trusted third party so as to result in non-universal about the presented protocol. In [13,14], Bi-directional authentication among nodes is presented. Peyravian mentions an authentication scheme based on Hash function [15]. Ma proposes a point-to-point authentication and secure transmission protocol [16] based on Hash functions and block cipher. A secure transmission method which is fit on the Internet of Things has been mentioned in [17]. The trusted third party is adopted while two parties are authenticating, therefore the scheme is not universal in terms of the complex web environment. In secure transmission model of Internet of Things, there is a common problem in the application, and there involves a variety of mixed-format electronic medical records and other patient data in Medical Internet of Things.

However, methods which we have discussed above cannot fit the field. When data is transferred, data attacking and data leaking leads our privacy information to be illegally obtained.

The rest of the paper is organized as follows. Section 2 describes the scheme model of the primary care. Section 3 provides our transmission protection scheme. Section 4 presents the results of our theoretical analysis. The last section concludes the paper and lays out future research directions.

3   Scheme Model in Primary Care

The Medical Internet of Things scheme is achieved in the community. Primary Care Architecture that describes the medical data sources and data transmission is presented in Fig. 1, and the slice model of medical data transmission that describes the transfer process of slicing data is showed in Fig. 2.

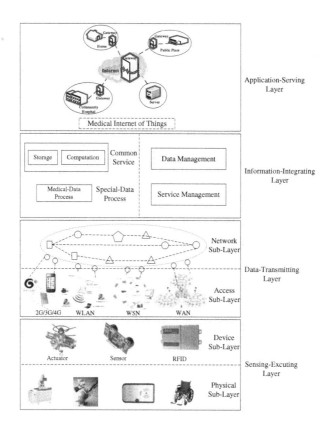

Fig. 1. Primary Care Architecture.

3.1   Primary Care Architecture

As we can see in Fig. 1, it describes an architecture of Primary Care. Data transmission integrates a variety of communication means. Sensors establish communication via Wireless self-organized network, and data in the gateway transmit through Wireless Local Area Network or mobile network.

Data from a sensor is sent toward the nearest gateway, and then the data is transmitted to the final community gateway. Connection is built between the community gateway and the database server through wireless network. In the end, the application database server provides the resolved data to users.

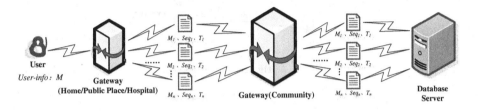

Fig. 2. Medical Data Multi-path Transmission Model.

3.2   Medical Data Multi-path Transmission Model

Medical data multi-path transmission model is painted in Fig. 2. We assume that the user's information is M, and the number of paths is n. Data is divided into n divisions when data arrives in the gateway. Each division which contains certain user information is transmitted to the database server.

4   Transmission Protection Scheme

4.1   Scheme Initialization

There needs to be an authentication with each other before the interaction between the Gateway-node and the Server. On the basis of the previous study about authentication protocol, a new Bi-directional password authentication method is showed as follows.

Gateway Node Registration

When G registers at S, G delivers the hash value of password PW_G to S, then S contrasts the hash value of password to dictionary to authenticate. Many Gateway-nodes' passwords constitute a password table. S generates a symmetric key K_{S-G}, and secretly informs G. Ultimately, both securely store K_{S-G} (Table 1).

Bi-directional Authentication Process

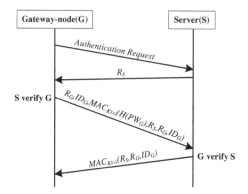

Fig. 3. Authentication Process.

Table 1. The Initial Conditions and Symbols

Number	Symbol	Definition
1	S	Server
2	G	Gateway-node
3	ID_G	Gateway-node Identification
4	PW_G	Gateway-node Password
5	$H(x)$	Strongly Non-collision Hash Function
6	R_G, R_S	Random number of Gateway-node and Server
7	K_{S-G}	Symmetric key for Gateway-node and Server
8	$MAC_{K_{S-G}}(M)$	The MAC value of M in the key K_{S-G}

4.2   Key-Agreement Mechanism

After completing Bi-directional authentication between G and S, they need to generate a shared key. Key generation and distribution process are elaborated as follows.

Step 1: A large prime number P is selected from G and S, and G is selected as a generator for the multiplicative group Z_P^*;

Step 2: G chooses a secret integer x:

$$1 \leq x \leq P - 2 \tag{1}$$

calculates $X = g^x mod\ P$, and sends X to S;

Step 3: S selects a secret integer y:

$$1 \leq y \leq P - 2 \tag{2}$$

calculates $Y = g^y \bmod P$, and sends Y to G;

Step 4: G calculates $K_G = Y^x \bmod P$, generates random number N_G, and sends $\{E_{K_G}(N_G), N_G\}$ to S;

Step 5: S calculates $K_S = X^y \bmod P, E_{K_G}(N_G)$ and generates N_S, and then sends $\{E_{K_S}(N_S), N_G, N_S\}$ to G;

Step 6: G receives and decrypts $\{E_{K_S}(N_S), N_G, N_S\}$, and then returns True to S, the two parties share the same key K_{S-G}, which is used to complete the key sharing.

4.3   Fragmented Multi-path Data Transmission

According to the mentioned above, G and S have accomplished Bi-directional authentication, and commonly share session key. To ensure the security of the data transmission process, G encrypts data by a shared key before data transmitting, and divides cipher-text into fragment to transfer. Multi-path data encryption and cipher-text transmission are described as follows.

Step 1: G uses the key K_{S-G} to encrypt the transmission data. Assuming that the data packet to be transmitted is M, the cipher-text is $C = E_{K_{S-G}}(M)$;

Step 2: C is divided into sub data packets $C_1, C_2, ..., C_n$. For every one of the sub data packets, we add a session number seq, sub-packet identification i and time stamp T_i to them,

$$m_i : \{C_i, s, i, T_i\}(1 \leq i \leq n); \tag{3}$$

Among them, the session number seq and sub packet identification i are used to be prevent replay attacking. $\{H_{K_{S-G}}(C_i, s, i, T_i)\}$ is calculated by $H(x)$, which is used to verify the message for receiver, and the message is transmitted on the each selected path.

$$S_i\{C_i, s, i, T_i, H_{K_{S-G}}(C_i, s, i, T_i)\}; \tag{4}$$

Step 3: Every packet of data received, S will authenticate the message according to the authentication code;

Step 4: When the server receives all of sub data packets which are sent from G, S will reorganize and decrypt the sub data packets to recover data packet M according to the sub-packet identification if the authentication is passed.

5   Security Analysis

5.1   Authentication

G has an authentication with S before data transmitting. If the two parties are not entirely passed the certification, S refuses to receive data in case of leaking

Table 2. Authentication security analysis

Security condition \ Scheme	Hwang[13]	Peyravian[14]	Wang[15]	Ma[16]	This paper
Prevent DoS attacking	-	Yes	-	Yes	Yes
Prevent replay attacking	-	Yes	Yes	Yes	Yes
Prevent dictionary attacking	Yes	Yes	Yes	-	Yes
Prevent Server forging	Yes	Yes	Yes	Yes	Yes
Prevent Gateway-node forging	Yes	-	Yes	Yes	Yes
No public key mechanism	Yes	Yes	-	Yes	Yes
Hash function	Yes	Yes	Yes	Yes	Yes
MAC function	-	-	-	Yes	Yes

data to fake nodes. Even if attackers steal the password table of S, they cannot crack the password because of the unidirectional characteristic of Hash function. Therefore, it can effectively ensure the identity authentication for S and G.

As shown in Table 2 where "Yes" represents that the security condition is met, from the implementation process, the above protocol takes advantage of Hash function and MAC to become more efficient than Wang's public key algorithm. MAC function is not referred in Peyravian's research, therefore Peyravian's protocol cannot prevent Gateway forging. MA's protocol cannot prevent dictionary attacking due to the data characteristic despite of various means of attacking.

5.2 Key Agreement

In order to prevent the attacker from forging new data to result in inconsistence while the two parties are exchanging information, here, we bring three-way handshake during the session so as to ensure the correctness of the final key-agreement. In Table 3, "Yes" represents that the security condition is met.

Table 3. Key-agreement security analysis

Security condition \ Scheme	Diffie-Hellman	Xie[17]	This paper
Prevent replay attacking	Yes	Yes	Yes
Forward security	Yes	Yes	Yes
Integrity attacking	Yes	Yes	Yes
Known key security	Yes	Yes	Yes
Prevent wiretap attacking	Yes	Yes	Yes
Prevent MITM attacking	-	Yes	Yes
Three-way handshake	-	-	Yes

6   Experimental Results

According to the security analysis, data packet will be transmitted to the Gateway Node after authenticating between Server and Gateway. Assuming that the probability of data stolen in single-path is $P(0 < P < 1)$, the probability in multi-path is P^n (n presents the number of paths). Again, we assume that P is 0.7, and then the maximum number of paths is 20, simulated by Matlab as shown in Figs. 4 and 5.

Assuming that the length of the communication link from terminal node to the server is L, and k nodes will be attacked by k attackers, then we conclude that the probability of effective node is $P = 1 - k/L$.

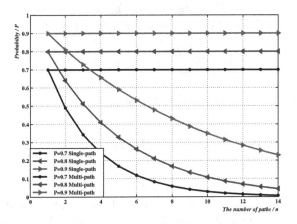

Fig. 4. Packet loss rate.

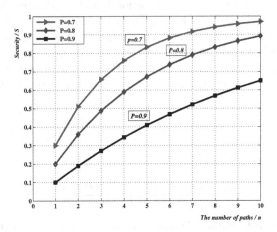

Fig. 5. Security of Communication Link.

Another assumption is that terminal node, server locates in the position 0 and $L + 1$, and the reliability of security is R.

$$A(1 \leq x \leq L - k) \tag{5}$$

represents that the node in the communication link is attacked by the first attacker. A' indicates that the first node which is attacked by the x_{th} node or the former node. $P(x)$ expresses that the node which is attacked locates at the x_{th} node in the communication link. G indicates that the identity of sender is guessed correctly by the first attacker and $P(G \mid A'_G)$ expresses the probability of attacking correctly.

According to the assumption above, the probability of the first node which is attacked by attacker locates at the x_{th} node in the communication link is $P(A_x) = P^{(x-1)}(1 - P)$. The probability of the first node which is attacked by attacker is in the position of first node or the latter node is

$$(1 - P) \sum_{i=1}^{L-k} P^{i-1} = 1 - P^{L-k} \tag{6}$$

Therefore we can conclude the probability of $P(G \mid A'_i)$ is

$$\frac{1 - P}{1 - (P^{L-k})} = \frac{1 - P}{1 - (P^{L-P})}, \tag{7}$$

We also assume that

$$LA_x(k \leq x \leq L) \tag{8}$$

represents that the last node which is attacked locates at the node in the communication link. LA'_x indicates that the first node which is attacked by the x_{th} node or the latter node. $P_{l(x)}$ expresses that the last node which is attacked locates at the x_{th} node in the communication link. G_l indicates that the identity of receiver is guessed correctly by the last attacker and $P_l(G \mid A'_G)$ expresses the probability of attacking correctly. According to the assumption above, the probability of the last node which is attacked by attacker locates at the x_{th} node in the communication link is $P(A_x) = (1 - P)P^{L-x}$. The probability of the last node which is attacked by attacker is in the position of n_{th} node or the latter node is

$$P(A_n^i) = (1 - P) \sum_{j=k}^{L} P^{L-j} \tag{9}$$

So we can also conclude the probability of $P(G_l \mid A'_n)$ is

$$\frac{1 - P}{1 - P^{L-k+1}} = \frac{1 - P}{1 - P^{L \times P + 1}} \tag{10}$$

To sum up from formulas (7) and (10), we can come to a decision that the reliability is

$$R = P(G \mid A'_i) \times P(G_l \mid M'_n) = \frac{(1 - P)^2}{(1 - P^{L \times P}) \times (1 - P^{L \times P + 1})} \tag{11}$$

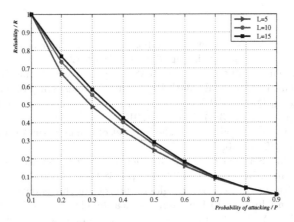

Fig. 6. The tendency of the reliability of communication link.

In order to reflect more directly the influence of the reliability about data transmitting with the impact of the length of the communication link and the probability of the node which is attacked, we have a simulation as shown in Fig. 6.

7   Conclusion

This paper takes the Medical Internet of Things as a standpoint, and aims at the security of data transmission. Then we summarize upon three aspects, authentication, communication key agreement and Multi-path security transmission. Considering of the security problems that might exist in the communication process, we improve the traditional key agreement algorithm to enhance the key negotiation security. Furthermore, we increase the multi-path transmission mechanism to become more difficult for attacker to obtain complete data without affecting Server data receiving. Finally, we analysis the security about the method inferred to the full text.

In a word, the security issue in Medical Internet of Things we discussed here is a part of the whole. And the protection of medical data storage, medical data privacy protection and other issues are also the research focus problems. The next step about our work is to explore and research for the protection of medical data storage.

Acknowledgements. The authors would like to thank our colleagues and students in Engineering Technology Research Center of Network and Information Security at Anhui Normal University, in particular, Yonglong Luo, Xuemei Qi and Yunxiang Sun. We thank National Natural Science Foundation of China under Grant No. 61572036, University Natural Science Research Project of Anhui Province under Grant No. KJ2014A084, Wuhu City Science and Technology Project under Grant No. 2014cxy04, and Anhui Normal University Postdoctoral Project under Grant No. 161-071214 for support of this research.

References

1. Gao, M., Zhang, Q., Ni, L., et al.: Cardiosentinal: a 24-hour heart care and monitoring system. J. Comput. Sci. Eng. **6**(1), 67–78 (2012)
2. Zhang, J., Chen, D., Zhao, J., et al.: RASS: a portable real-time automatic sleep scoring system. In: 2013 IEEE 34th Real-Time Systems Symposium (2012)
3. Mni, L., Zhang, Q., Tan, H.Y., et al.: Smart healthcare: from IoT to cloud computing. Scientia Sinica **43**(4), 515–528 (2013)
4. Atzori, L., Iera, A., Morabito, G.: The internet of things: a survey. Comput. Netw. **54**(15), 2787–2805 (2010)
5. Medaglia, C.M., Serbanati, A.: An overview of privacy and security issues in the internet of things. In: Internet of Things, pp. 389–395 (2009)
6. Ning, H.S., Xu, Q.Y.: Research on global internet of things developments and its lonstruction in China. Acta Electronica sinica **38**(11), 2590–2599 (2010)
7. Wu, Z.Q., Zhou, Y.W., Ma, J.F.: A secure transmission model for internet of things. Chin. J. Comput. **34**(8), 1351–1364 (2011)
8. Du, W., Deng, J., Han, Y.S.: A pairwise key pre-distribution scheme for wireless sensor networks. In: CCS 2003 Proceedings of the 10th ACM Conference on Computer and Communications Security, vol. 8, issue: 2, pp. 42–51 (2003)
9. Song, Z., Zhang, Y., Wu, C.: A reliable transmission scheme for security and protection system based on Internet of Things. In: International Conference on Communication Technology & Application IET Digital Library (2011)
10. Lamport, L.: Password authentication with insecure communication. Commun. ACM **24**(11), 770–772 (1981)
11. Kothmayr, T., Schmitt, C., Hu, W., et al.: A DTLS based end-to-end security architecture for the Internet of Things with two-way authentication. In: IEEE Conference on Local Computer Networks Workshops, vol. 90, issue: 1, pp. 956–963 (2012)
12. Groce, A., Katz, J.: A new framework for efficient password-based authenticated key exchange. In: Proceedings of the 17th ACM Conference on Computer and Communications Security ACM (2010)
13. Hwang, J., Yeh, T.: Improvement on Peyravian-Zunics password authentication schemes. IEICE Trans. Commun. **85**(4), 823–825 (2002)
14. Wang, B., Zhang, H., Wang, Z., et al.: A secure mutual password authentication scheme with user anonymity. Geomatics Inf. Sci. Wuhan Univ. **33**(10), 1073–1075 (2008)
15. Peyravian, M., Jeffries, C.: Secure remote user access over insecure networks. Comput. Commun. **29**(5), 660–667 (2006)
16. Ma, W.J.: Research and Application on Security Authentication Technologies in Internet of Things. Shandong University (2011)
17. Xie, W.J.: A Secure Communication Scheme based on Multipath Transportation for the Internet of Things. South China University of Technology (2013)

Robust Topology and Chaos Characteristic of Complex Wireless Sensor Network

Changjian Deng[1(✉)] and Heng Zhang[2]

[1] Department of Control Engineering, Chengdu University
of Information Technology, Chengdu 610225, China
Chengli_dcj@163.com
[2] School of Computer and Information Science,
Southwest University, Chongqing 400715, China

Abstract. Wireless sensor networks have found their chaos characteristics in their applications, for example, in network coding, synchronizing, network security communicating, and so on. To analyze the robustness of WSN under complex status, the nonlinear dynamic equation of network connectivity changed by attack is proposed, and then the bifurcation of network dynamic connection topology is analyzed in the paper. It shows that there is a limit to predict or control the network topology in the global (or top to down design), but it can be partly overcome by inverse design (or down to top design). This analytical finding is confirmed by numerical simulations; meanwhile it presents an inverse design method that is capable of stabilizing network topology under attack.

Keywords: Chaos · Bifurcation · Robust topology · Complex wireless sensor network · Nonlinear dynamic equation

1 Introduction

Chaotic system is sensitive with its initial values; it has the statistic characteristic of white noise, ergodic characteristic of chaotic sequences, very complex fractal structure; and meanwhile is unpredictable.

The nodes, relays, routes in WSN may lose their functions when they are to be attacked. The connection status of network is very complex and unpredictable when the interferences are unknown. Most median or large scale networks have complex dynamic architecture due to their large network capacity, scalability, and mobility. For example, the robust topology is needed when network is under environment stressed, load changed and nodes to be attacked. The related research includes: Helmy studies the small-world network effects in wireless sensor network [1]; other researchers have studied the graph theory [2, 3], scaling theory [4, 5] of WSN.

The interest findings are that the network is regarded as dynamic nonlinear system, and can be classified as three kinds of related chaos. The first kind of chaos comes from the asymptotic integration of Lienard equation [6, 7]. The second kind of chaos comes from the modification of harmonic linearization and describing function (when the generalized Routh–Hurwitz conditions are satisfied) [8, 9]. The third kind of chaos comes from the strange attractors [10, 11].

© ICST Institute for Computer Sciences, Social Informatics and Telecommunications Engineering 2016
J. Wan et al. (Eds.): Industrial IoT 2016, LNICST 173, pp. 155–165, 2016.
DOI: 10.1007/978-3-319-44350-8_16

In researches of robust topology of complex sensor network, paper [12] studies detecting topological holes in WSN with no localization information, for example, using a distributed scheme based on the communication topology graph. Paper [13, 14] use an algebraic topological method to detect single overlay coverage holes without coordinates based on homology theory. Paper [15] presents a heuristic method to detect holes based on the topology of the communication graph. Paper [16] presents a coordinate-free method to identify boundaries in WSNs. Paper [17] proposes a self-organization framework based on topological considerations and geometric packing arguments, to determine the boundary nodes and the topology of the whole network.

This paper focuses on robust topology and chaos characteristic of wireless sensor networks under complex status. The contributions of paper include:

(1) The paper presents dynamic nonlinear equations of network topology when they are to be attacked or they lose functions themselves in networks.
(2) The paper analyzes the chaos and Bifurcation of topology of wireless sensor networks.
(3) The paper gives the design method of robust topology of WSN under complex status.

The rest of the paper is organized as follows: Sect. 2 presents the fundament of Bifurcation equation; Sect. 3 studies the dynamic nonlinear equations and network topologies with some lose function nodes; Sect. 4 introduces the robust topology of wireless sensor network; and Sect. 5 is the conclusion of the paper.

2   Bifurcation and Robust Topology

The median and large scale wireless equipment condition monitoring system is a dynamic nonlinear system; its data stream works like Geophysical fluid, it has the characteristics of dissipative system.

Bifurcation theory studies the changes in the qualitative or topological structure of a given system. It has almost the opposite meaning of robustness, a bifurcation occurs when a small smooth change made to the parameter values (the bifurcation parameters) of a system causes a sudden 'qualitative' or topological change in its behavior. Table 1 shows different bifurcations occur in both continuous systems (described by ordinary differential equations-ODEs, delay differential equation-DDEs or Partial Differential Equations-PDEs), and discrete systems (described by maps).

More general, consider the continuous dynamical system described by the ODE.

$$\dot{x} = f(x, \mu); f : \Re^n \times \Re \to \Re^n \tag{1}$$

In Eq. (1), x is state variable, μ is control variable, $\Re$ represents that the variable is in one dimensional real space, $\Re^n$ represents that the variable is in n dimensional real space.

A local bifurcation occurs at (x_0, μ_0) if the Jacobian matrix df_{x_0, μ_0} has an eigenvalue with zero real part.

Table 1. Bifurcation types and their discription

Type	Continuous systems: differential equation ('x' is state variable, 'μ' is control variable.)	Discrete systems: difference equation ('x' is state variable, 'μ' is control variable.)
Pitchfork bifurcation	$\dot{x} = \mu x - x^3$	$x_{n+1} = (\mu+1)x_n - x_n^3$
Tangent bifurcation	$\dot{x} = \mu - x^2$	$x_{n+1} = \mu + x_n - x_n^2$
Trancritical bifurcation	$\dot{x} = \mu x - x^2$	$x_{n+1} = (\mu+1)x_n - x_n^2$
A hysteresis bifurcation	$\dot{x} = \mu + \gamma x - x^3$	$x_{n+1} = \mu + (\gamma+1)x_n - x_n^3$
Symmetry breaking bifurcation	$\dot{x} = \mu + \gamma x - x^3$	$x_{n+1} = \mu + (\gamma+1)x_n - x_n^3$
Horf bifurcation	$\dfrac{dx}{dt} = -\gamma \cdot y + x[\mu - (x^2 + y^2)]$ $\dfrac{dy}{dt} = \gamma \cdot x + y[\mu - (x^2 + y^2)]$	$x_{n+1} = y_n$ $y_{n+1} = \gamma \cdot y_n(1 - x_n)$
Period doubling bifurcation	$\dfrac{dx}{dt} = -y - z$ $\dfrac{dy}{dt} = x + 0.2y$ $\dfrac{dz}{dt} = 0.2 + xz - \mu z$	$x_{n+1} = 4\lambda x_n(1 - x_n)$

3   Dynamic Nonlinear Equations and Lose Function Node Network Topology

This section has two parts: part one discusses the nonlinear dynamic models of complex wireless sensor networks; part two analyzes the network connection topology which has failure nodes.

3.1   The Nonlinear Wireless Sensor Networks Transmitting Model

The data transmitting equation of wireless sensor networks can be described as Eq. (2).

$$\sum_{j=1}^{K} \sum_{i=1}^{k} X_j(x_i) = Y \qquad (2)$$

In Eq. (2), there is the number of K distributed wireless nodes transmitting data, and X_j represents any one of K nodes. In a certain period, every one of X_j has k data sequence $x_1, ..x_k$. And Y is the transmitted data sum of a determined time period.

From the view of Eq. (2), the data stream of network is transmitting and receiving data sequence from different nodes. An ideal wireless monitoring system should have high accuracy, dynamic network topology, less missing data, and so on. Because the communication environment can influence data transmitting process, and this interferences factor is random, then the real receiving data can be supposed to be a function of Eq. (2). Then we have constraint Eq. (3).

$$\sum_{j=1}^{K}\sum_{i=1}^{k} X_j(x_i) - f\left(\sum_{j=1}^{K}\sum_{i=1}^{k} X_j(x_i)\right) < y_{cs} \tag{3}$$

Here y_{cs} is limit of transmitting data error or missing data number.

In multi hop communications of wireless sensor networks, the transmitting node X_j and data sequence x_i in their transmitting process should change, miss, disorder, and delay.

Meanwhile nodes of wireless sensor networks have limited resource and calculate ability, the Eq. (3) can be rewritten as (4). And Λ is influence factor, F is influence function of corresponding node; p is influence possibility of the every different data sequence. Paper proposes that the received data is often to be transmitted in good communication path, and then obtain Eq. (4). As in a certain period, the transmitted data sum may be different with different link path and different start time, so $Y_{i'j'}$ in Eq. (4) mean first node j' and first data sequence number i'. So, more generally, Y in Eq. (2) can be replaced by Y_{ij}.

$$\begin{cases} F(Y_{i'j'}) = \sum_{j=j'}^{K+j'}\sum_{i=i'}^{i'+k'} \Lambda \cdot X_j(x_i) \\ \prod_{j'=1,..k} p(F(Y_{i'j'})) = \min \end{cases} \tag{4}$$

So paper considers the link chain of network, especially the relationship of next link node communication status variable x_{n+1} with current communicate node status variable x_n.

$$x_{n+1} = f(x_n, u(m), \theta) \tag{5}$$

In Eq. (5), $x \in \Re^n$, its initiate value is x_1, control variation $u \in \Re^m$, and θ is uncertain parameter.

It is obvious that dynamic network topology is complex, and its delay, packet loss rate make (5) have characteristic of nonlinear equation. Without loss of generality, the paper will analyze its dynamic process using the geometric method.

3.2   The Analysis of Network Connection Topology of Failure Nodes Based on the Differential Equation

If a full function node (for example relay or route node) lose its function abruptly, it should directly influence the topology of WSN, and then influence the correctness of monitoring data.

Table 2. Some bifurcation examples of the network connection status of failure nodes

topology	network connection status	Control parameters μ and cannot connect nodes in different status x_i	difference equation of network connect status
bidirectio nal ring		$\mu = 1;$ $x_1 = -1; \rightarrow x_2 = -1;$ $x_3 = -1;$ $\rightarrow$ .....Tangent bifurcation	$\dot{x} = \mu - x^2$ or $x_{n+1} = \mu +$
binary tree		$\mu = 0;$ $x_1 = -1; \rightarrow x_2 = -2;$ $x_3 = -6;$ $\rightarrow$ .....Tangent bifurcation	$\dot{x} = \mu - x^2$ or $x_{n+1} = \mu +$
Two bidirectio nal ring having two sub network		$\mu = 4;$ $x_1 = -2; \rightarrow x_2 = -2;$ $x_3 = -2;$ $\rightarrow$ .....Tangent bifurcation	$\dot{x} = \mu - x^2$ or $x_{n+1} = \mu +$
Three bidirectio nal ring having two sub network		$\mu = 9;$ $x_1 = -3; \rightarrow x_2 = -3;$ $x_3 = -3;$ $\rightarrow$ .....Tangent bifurcation	$\dot{x} = \mu - x^2$ or $x_{n+1} = \mu + x$

Then, the paper discusses the bifurcation of network. First, x_1 is defined as the number of nodes that cannot be connected in initiate status (or lowest layer), in next status (or next layer) the number of nodes that cannot be connected is defined as $x_2, \ldots,$ in the end status (sink layer) this is defined as x_n. And then paper defines the control parameter is μ (its physical meaning is characteristics parameter of network topology). Table 2 discusses some examples of failure point network connection status and their differential equations.

Notice, the bidirectional ring can form spiral, and the MESH network can form bidirectional ring, they all have the self like structure. This is why the paper uses the Tangent bifurcation to describe the failure point network connection equation.

4 Robust Topology in Complex Wireless Sensor Network Having Disconnected Nodes

4.1 Chaos in Complex Topology Network Communication

At the beginning of this section, paper presents some definitions.

Definition 1: network connection missing degree n. It means in a topology, if the number of nodes which have lost their function (can not communicate to other nodes) is n, then the number of nodes in this network topology that cannot transmit their test data to the data center is $n + 1$, then define the network connection missing degree is n. For example, in the figure, two bidirectional ring having two sub network of Table 1, although, physically only one node lose function, but in logically, same network control parameter make self like structure lose simultaneously (for example, in one work mode: the allocated time slot is same for subnetwork to work simultaneously in large TDMA network or other network).

Definition 2: the folded node, the independent node, the terminal node. The degree (or valency) of a vertex of a graph is the number of edges incident to the vertex. If in degree and out degree of a network is not equal, and the network connection missing degree is 1, the node is folded node.

If in degree and out degree of a network is not equal, and the network connection missing degree is 1, the node is independent node.

Have only in degree, the node is terminal node.

Theory 1: if the network connection missing degree were 1, the transform change of topology of this network can form 'smale horseshoe'. It has characteristics of local unpredictability by different initial value.

Investigate a network which the protocol of communication is Zigbee, wireless HART and so on; the topology of network is formed as below:

- Ideal MESH construction is made by all independent nodes, for example, the full function nodes;
- Tree cluster network is often made by relay nodes, terminal node and sink node; for example, ZIGBEE

- if a network have arbitrarily topology, for example two bidirectional ring, MESH, and so on, then the network is complex network.

Table 3 shows if the network connection missing degree is 1, then the topology of two bidirectional ring can be stretched and folder, so the two bidirectional ring can do 'smale horse shoe transformation', then the communication of complex wireless sensor network having disconnected nodes may have chaos phenomenon.

In detail, a complex wireless sensor network which has disconnected nodes, the communications of nodes have the following properties:

- *it is sensitive to initial conditions; the initial disconnect nodes in different nodes have large difference. For example, in the network missing connection degree is 1, if there is already have one node lose function, then, the next place of lose function node can have very difference and unprediction communication status, so is next node, and so on.*
- *it is topologically mixing; and*
- *it has dense periodic orbits. These two properties is complex network communication native properties, for the orbit or communication path is very complex, it can be topologically mixing, and has dense periodic orbits.*

Table 3. The network topology tranform with a network missing connection degree is 1

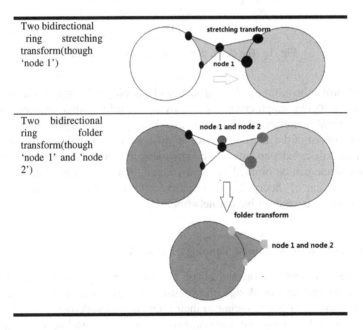

4.2   Deployment and Communication in Complex Network

Theory 2: The binary tree topology which is constrained by the communication distance cannot deploy arbitrary shape of 2D in large scale network.

Prove of theory 2: Here use the method of proof by contradiction. And propose, the deployment has no other information.

First, select an infinitesimal (for example square) of a network, if the theory 2 were correct, then in Fig. 1, the number of nodes in a square should be 5, and if four vertex add center point, the edge 'l1' and 'l2' (link path from node 0 to node 2, and link path from node 1 to node 2) break binary tree topology. This is contradiction with propose.

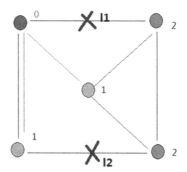

Fig. 1. The demo picture of link path of square

Second, prove the arbitrary shape of 2D can be deployed by square infinitesimal. It is obvious that regular graph can be formed by square infinitesimal. So here only need considers the odd boundary of arbitrary shape. As a special example, suppose wireless sensor node consist of the auxiliary sensor module and the communication module, and the sensor module is independent with communication module. Then the odd boundary of arbitrary shape can be omitted by deploying auxiliary sensor module in this area.

As a deduction of theory 2, the communication distance determined cluster-tree topology network is random branch network.

4.3   Inverse Design of Robust Topology Network

As there are many bifurcation and chaos phenomenon in complex wireless sensor network. Then one cannot design the robust complex wireless sensor network in general or from top to down, owing to their unpredicted properties.

To design a robust topology network, here reference the local network design. For example, when we design network local Ethernet, we do not concern the whole world internet network, we only concern the local network. We design a network from down to top, or do inverse design.

From the view of this, the robust topology of network is in a very small local area, to form bidirectional ring, for example to form redundancy ring in cluster head that can be robust topology in complex wireless sensor network.

5   Simulation and Results

5.1   Phase Plot of Communication Having Random Disconnected Nodes

In Fig. 2, the communication area is changed when disconnected node changed, the communication plot with the lose node changed, but as circle, the whole area can all be covered (except the disconnected node itself). This is a explanation of network missing connection degree.

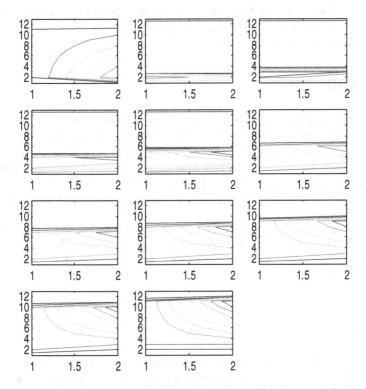

Fig. 2. The connection status of 12 nodes bidirectional ring which have a random lose funtion node

5.2   Eddy Phase Plot of Four Connection Nodes

In Fig. 3, there are four bidirectional nodes in topology, the phase plot of its communication area. It is obvious, the ring phase interspaces have cross connected phenomenon, have circulation pattern.

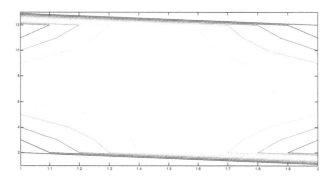

Fig. 3. The connection status of 4 bidirectional ring nodes

In Fig. 3, there are four bidirectional nodes in topology, the phase plot of its communication area. It is obvious, the ring phase interspaces have cross connected phenomenon, have circulation pattern.

6 Conclusion

To analyze the robustness of WSN under complex status, the nonlinear dynamic equation of connectivity network topology is introduced, and then the fractal of dynamic wireless sensor network topology is analyzed in the paper.

The analysis shows that there is a limit that to predict or control the network topology in the global (or top to down), but it can be partly overcome by inverse design (or down to top). This analytical finding is confirmed by numerical simulations.

Acknowledgment. This article is supported by "China-Canada Joint Research Project" (project number: 2009DFA12100) and Major Project of Education Department in Sichuan (14ZA0172). For support, contributions to discuss we would like to thank the rest of MCC (mobile computer center) of UESTC.

References

1. Helmy, A.: Small worlds in wireless networks. IEEE Commun. Lett. **7**(10), 490–492 (2003)
2. Donev, A., et al.: Improving the density of jammed disordered packings using ellipsoids. Science **303**, 990–993 (2004)
3. Franceschetti, M., Cook, M., Bruck, J.: A geometric theorem for network design. IEEE Trans. Comput. **53**(4), 483–489 (2004)
4. Barenblatt, G.I.: Scaling. Cambridge University Press, Cambridge (2003)
5. Watts, D.J., Strogatz, S.H.: Collective dynamics of 'small-world' networks". Nature **393**(4), 440–442 (1998)
6. Carmona, P., Franco, D.: Control of chaotic behaviour and prevention of extinction using constant proportional feedback. Nonlin. Anal. Real World Appl. **12**, 3719–3726 (2011)

7. Eskola, H.T.M., Parvinen, K.: The Allee effect in mechanistic models based on inter-individual interaction processes. Bull. Math. Biol. **72**, 184–207 (2010)

8. Hilker, F.M., Westerhoff, F.H.: Preventing extinction and outbreaks in chaotic populations. Am. Nat. **170**, 232–241 (2007)

9. Liz, E., Tkachenko, V., Trofimchuk, S.: A global stability criterion for scalar functional differential equations. SIAM J. Math. Anal. **35**, 596–622 (2003)

10. Liz, E.: How to control chaotic behaviour and population size with proportional feedback. Phys. Lett. A **374**, 725–728 (2010)

11. Liz, E.: Complex dynamics of survival and extinction in simple population models with harvesting. Theor. Ecol. **3**, 209–221 (2010)

12. Kun, B., Kun, T., Naijie, G., Wan, L.D., Xiaohu, L.: Topological hole detection in sensor networks with cooperative neighbors. In: Proceedings of the International Conference Systems and Networks Communications (ICSN 2006) (2006)

13. Ghrist, R., Muhammad, A.: Coverage and hole-detection in sensor networks via homology. In: Proceedings of the Fourth International Symposium Information Processing in Sensor Networks (IPSN 2005), pp. 254–260 (2005)

14. De Silva, V., Ghrist, R., Muhammad, A.: Blind swarms for coverage in 2-D. In: Proceedings of the Robotics: Science and Systems, pp. 335–342 (2005)

15. Stefan, F.: Topological hole detection in wireless sensor networks and its applications. In: Proceedings of the Joint Workshop on Foundations of Mobile Computing, pp. 44–53 (2005)

16. Fekete, S.P., Kröller, A., Pfisterer, D., Fischer, S., Buschmann, C.: Neighborhood-based topology recognition in sensor networks. In: Nikoletseas, S.E., Rolim, J.D. (eds.) ALGOSENSORS 2004. LNCS, vol. 3121, pp. 123–136. Springer, Heidelberg (2004)

17. Fekete, S.P., Kaufmann, M., Kr€oller, A., Lehmann, N.: A new approach for boundary recognition in geometric sensor networks. In: Proceedings of the 17th Canadian Conference on Computational Geometry, pp. 82–85 (2005)

A Novel Algorithm for Detecting Social Clusters and Hierarchical Structure in Industrial IoT

Jiming Luo, Kai Lin$^{(\boxtimes)}$, and Wenjian Wang

School of Computer Science and Technology, Dalian University of Technology,
Dalian, China
logicluo@foxmail.com, link@dlut.edu.cn, dlutwwj0boa@mail.dlut.edu.cn

Abstract. The rapid development of IoT has brought life around us with tremendous impact. Especially, industrial IoT as a new research hotspot, has been attracting extensive concern from industry and academia, facilitating many technologies and application in industrial IoT. However, taking full advantage of a large number of resources in industrial IoT is a challenging task. In this article, we present an efficient mobile social cluster algorithm (OMSC) to detect the potential social relationships among mobile devices in industrial IoT. It can discover the overlapping cluster and hierarchical structure in near-line time. We implement this algorithm in the Java Platform and validate the OMSC in synthetic networks and real-world network datasets. The experimental results demonstrate that the presented OMSC algorithm has high performance.

Keywords: IoT · Industrial cloud · Social cluster · Hierarchical structure · Mobile device

1 Introduction

With the rapid development of Internet of Thing (IoT), a splendid Internet blueprint containing all kinds of resources and services appears in front of people [1]. IoT brings economies and societies around us tremendous impact. The obvious feature of IoT is mainly the omnipresent information acquisition and ubiquitous information processing. The intellectualization progress of various field gains the effective pushing from IoT. Especially, IoT has been accelerated the development of next industrial revolution "Industry 4.0" [2]. Industrial Internet of Things (IIoT), as a remarkable feature of "Industry 4.0", is expected to provide high efficient schemes for many existing industrial fields such as manufacturing fields and transportation fields. Much concern from industry and academia around the world have been put into Industrial IoT, which greatly facilitates the progress of Industrial Internet of Things (IIoT).

With the gradually increasing number of mobile devices in industrial IoT, the way of communication among people has a dramatic changing. Worldwide shipments of tablets and mobile phones increasing to 2.072 billion units which

© ICST Institute for Computer Sciences, Social Informatics and Telecommunications Engineering 2016
J. Wan et al. (Eds.): Industrial IoT 2016, LNICST 173, pp. 166–178, 2016.
DOI: 10.1007/978-3-319-44350-8_17

is acceding PC-desk based and notebook 315,229 units in 2013 and also 37 % of PC users have switched to cellphone or tablet to surf the Internet or play games [3]. However, how to effectively utilize the resources among the mobile devices has certain challenge.

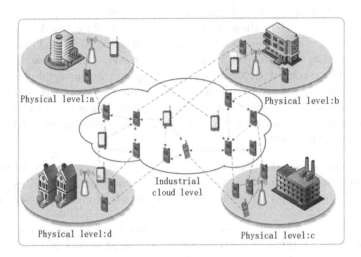

Fig. 1. An illustration of mobile device relationships in the industrial cloud level and physical level.

In industrial cloud applications where the data is shared by multiple mobile users, it is essential to provide consistency among mobile users by means of synchronization algorithms. In particular, if the data is frequently updated and the number of mobile users sharing the data is large, the synchronization traffic can significantly increase. Virtualisation of industrial cloud networks can be seen as a way to substantially reduce the complexity of processes. Due to the limitations of mobile devices, using mobile databases on mobile devices encounters certain scalability issues in mobile data accesses and storage. One alternative approach is to use databases based on industrial cloud to support mobile users and applications on mobile devices. To support the fast increasing needs of mobile data accesses and mobile computing services, social cluster can play an important role in industrial IoT.

Social cluster discovering is an essential tool to understand the architecture of industrial IoT, which gives a new perspective to create applications. In the cluster, as illustrated in Fig. 1, a key observation is that mobile devices are coupled not only in the physical level owing to the physical relationship, but also in industrial cloud level due to the social ties among them [4]. In the physical level, mobile devices may change their location in a short time, but the social relationships are not changed. Mobile devices in the same cluster usually have the similar function or characteristic in industrial IoT. For this reason, cluster in industrial IoT can be seen as a way of tackling challenges.

Numerous techniques have been developed for social cluster detection in industrial IoT. However, most of them only considered one mobile device belongs to one social cluster [5,6], or hierarchical and non-overlapping clusters [7], or non-overlapping and overlapping clusters [8–10]. Even though some solutions can detect hierarchical and overlapping clusters in industrial IoT, they need to acquire the whole networks information, and the computational complexity of them is high which makes them unsuitable for real-world networks.

In this paper, we are based on the observation of industrial IoT, and propose a novel algorithm to detect the mobile social cluster. The proposed algorithm can find the hierarchical structure and overlapping mobile social clusters with reasonable time and space complexity. In addition, the proposed method takes full advantages of local resources in industrial IoT and there is no need to acquire the whole network information. To summarize, the contributions of our paper are summarized as follows:

1. We present an overlapping mobile social cluster algorithm (OMSC), the thought of which is detecting the relationship among mobile devices in industrial IoT. It can find the overlapping mobile social cluster in nearly-linear time.
2. We also design a mobile hierarchical structure method, which makes use of the results of OMSC and illustrates the relation of mobile social clusters.
3. To validate the efficiency and effectiveness of the proposed method OMSC, we have implemented our algorithm in the Java Platform and evaluated it in synthetic and real-world network datasets.

The remainder of this paper is organized as follows. Section 2 presents some related works. In Sect. 3, we describe the cluster method in industrial cloud networks. Experiments and results are presented in Sect. 4. In the end, we draw our conclusions in Sect. 5.

2   Related Work

In industrial IoT, social clusters are the essential local structures, which involve a group of mobile devices that intercommunicate with each other densely but communicate to the others sparsely. A large number of mobile social cluster discovery methods have been proposed, especially in the past decade, which fall in two broad categories: (1) matrix transformation; (2) non-matrix transformation.

As for matrix transformation mobile social clusters detection methods, I. Psorakis et al. [11] propose a probabilistic clusters detection approach where a Bayesian non-negative matrix factorization model is utilized to form an industrial cloud networks through extracting overlapping modules. The integration of soft-partitioning solutions, allocation of node concernment scores for modules, and an intuitive basis is achieved in this scheme. Based on the matrix factorization approach, Y. Zhang et al. [12] present a bounded nonnegative matrix tri-factorization (BNMTF) method, which can detect the overlapping social clusters with reasonable time complexity.

On the other hand, U.N. Raghvan et al. [13] introduce a simple label propagation algorithm in which the network structure is viewed as the only guidance. Moreover, in the algorithm, a predefined objective function and the preferential information of the cluster don't need optimizing. In the process of the algorithm, a unique label is used to initialize every mobile device, and then each mobile device receives the label coming from its neighbors. In this iteration, the groups of mobile devices that densely connected each another reach a consensus on a unique label to form clusters. To contain the information about more than one cluster, G. Steve extends the label and propagation step: every node can currently belong to v clusters, where v is the number of clusters [14]. Z. Zhang et al. investigate an functional and efficient cluster detecting algorithm MOHHC, where the overlapping and hierarchical organization in industrial IoT can simultaneously is discovered. In the algorithm, all maximal cliques are first extracted from the original Industrial IoT, and these extracted maximal cliques are merged into a dendrogram based on the aggregative framework from MOHCC. Finally, the dendrogram is cut through and achieve a network partition by using maximum extended partition density [15].

3 Social Clusters Detection in Industrial Cloud Network

This section presents a novel overlapping and hierarchical mobile social clusters detection (OMSC) algorithm for industrial IoT. Social cluster is an important characteristic of Industrial IoT. Although the concept of mobile social clusters dose not have clear definition, most industrial IoT emerges cluster structures. For instance, a group of mobile devices densely intercommunicate and sparsely communicate with others in industrial IoT. Furthermore, mobile cloud networks also contain some hierarchical organizations. As shown in Fig. 2, cluster 1 and cluster 2 contain some mobile devices and they are components of cluster a. Cluster a and cluster b belong to the cluster A. The proposed algorithm include two parts: overlapping mobile clusters detection (OMSC) and hierarchical structures detection. To describe the algorithm more clearly, we use the vertex or node to replace the mobile device and the edge to replace the relationship of two different mobile devices.

3.1 Overlapping Mobile Clusters Detection

According to the potential social relationship in industrial IoT, we define a cluster set $C(c_a, c_b, ..., C_\varepsilon)$ which represents social clusters result of industrial IoT. Based on the analysis of social relationship, all the mobile devices in industrial IoT can be divided into clusters. We also define a vector P_ξ^λ for node ξ (λ represents the iterations which will be described detailly later), which storages some key-value pairs (*label, proportion*). In the key-value pair, the *label* represents the node (mobile device) belonging to the cluster (C_{label}) and the *proportion* is the possibility of the node belonging to the cluster (C_{label}). When the algorithm runs over, the remains of labels in the vector P_ξ^λ denote the node should be divided to which clusters.

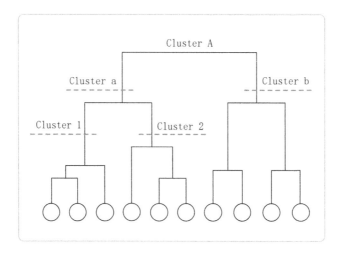

Fig. 2. An example of hierarchical structure in the industrial cloud networks.

Based on our observation, we find that a set of nodes in a complete subgraph are more likely to belong to a same cluster. So we come up with a new concept named primitive cluster (PC) in industrial IoT. Each pair of nodes in the PC is connected with each other and every nodes have the maximum number of the same neighbors existed in the same PC, and all the number of same neighbors should be equal to or less than this maximum value. The first step of the proposed overlapping mobile social clusters discovers the algorithm is to extract PCs from the industrial IoT.

In each node, a sequence of labels are maintained which can be broadcasted to its neighbors. We define n $1 \times n$ vectors P_ξ^λ (n is the number of nodes in the industrial cloud networks). For each key-value pair $(label, proporation)$ in the vector P_ξ^λ, we denote $P_\xi^\lambda(label) = proporation$ for convenience. In every broadcasts, each neighbor of node ξ send its vector to node ξ. Then, node ξ's vector will be computed by following equation:

$$P_\xi^{\lambda'+1}(label) = \sum_{i \in Nb(\xi)} \frac{P_\xi(label)}{k_\xi}, \quad \forall label \in L \tag{1}$$

where $Nb(\xi)$ is the neighbor set of node ξ and $k_\xi = |Nb(\xi)|$ is the number of neighbors. $L = \{a, b, c, ...\}$ (same as node id's) and $|L| = n$. As for some labels in the node ξ's neighbors not in ξ, we add these labels and their proportions on node ξ's vector directly. And we normalized $P_\xi^{\lambda'+1}$ using follow equation so that the proportions in $P_\xi^{\lambda'+1}$ sum to 1:

$$P_\xi^{\lambda''+1}(label) = \frac{P_\xi^{\lambda'+1}(label)}{\sum_{label \in P_\xi^{\lambda'+1}} P_\xi^{\lambda'+1}(label)} \tag{2}$$

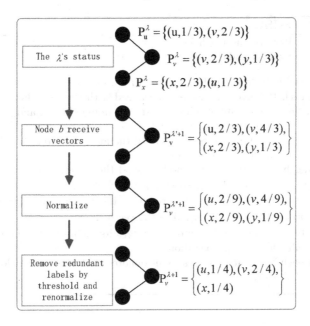

Fig. 3. Node v's vector changes from λ to $\lambda + 1$ iteration.

After normalizing the vector $P_\xi^{\lambda'+1}$, there are some redundant key-value pairs in the $P_\xi^{\lambda''+1}$. To solve this problem, we set a threshold ψ which value is initialized at the begin of algorithm. At each propagation step, we remove these key-value pairs in which the proportion is less than the threshold.

When the redundant key-value pairs are removed, the proportions in the vector $P_\xi^{\lambda''+1}$ may not satisfy the condition that the proportions in the vector $P_\xi^{\lambda''+1}$ are sum to 1. So we should use the Eq. 2 to re-normalize the vector. Finally, the step of propagation is finished completely.

$$P_\xi^{\lambda+1}\,(label) = P_\xi^{\lambda'''+1}\,(label) \tag{3}$$

There is an problem that when the algorithm stop or what is the algorithm terminate condition. In [13], they have been proved that if the number of iterations is larger than twenty, the algorithm has high performance. So when it run twenty times, the algorithm will be stopped.

To deeply comprehend the proposed algorithm, there is an example to explain it. As shown in Fig. 3, node v has two neighbors which denote u and x respectively. At the λ's iteration, $\mathrm{P}_u^\lambda = \{(u, 1/3), (v, 2/3)\}$, $\mathrm{P}_v^\lambda = \{(v, 1/3), (y, 2/3)\}$ and $\mathrm{P}_x^\lambda = \{(x, 2/3), (u, 1/3)\}$ (u, v, x and y, are *labels*). In the next iteration, the neighbors of node v send their vectors to node v, and node v use the Eq. 1 to compute its vector $P_v^{\lambda'+1}$, and the result is $P_v^{\lambda'+1} = \{(u, 2/3), (v, 4/3), (x, 2/3), (y, 1/3)\}$. Because the proportions in each vector sum to 1, we use equation to normalize the vector $P_v^{\lambda'+1}$, and it is

---

Algorithm 1. OMSC

T: the maximum iteration user defines.
ψ: threshold to cut off redundant labels.

1. First, extract PC from the industrial cloud networks.
2. Second, in each PC, every nodes are initialized with the same label and the rest of nodes in the network are also initialized with unique labels, and the proportion of each label in the vector is initialized 1.
3. Then, the following steps are repeated until the maximum iteration T is reached:
   (a) Selected one node v as a listener.
   (b) Each neighbor of the listener node v send their vector to node. Node v first compute $P_v^{\lambda'+1}$, and then normalize it to $P_v^{\lambda''+1}$.
   (c) Using the threshold ψ to remove the redundant key-value pairs $(label, propotion)$ and re-normalize the vector to $P_v^{\lambda'''+1}$.
   (d) The listener node v adds vector $P_v^{\lambda'''+1}$ to its memory and the vector as the initial value for next propagation.
4. Finally, in each node, its vector's label denotes that the node belong to which mobile social cluster.

---

changed to $P_v^{\lambda''+1} = \{(u, 2/9), (v, 4/9), (x, 2/9), (y, 2/9)\}$. The threshold is $1/9$ to cut off redundant labels, and the pair of $(y, 1/9)$ will be removed from the vector $P_v^{\lambda''+1}$. Then we re-normalize the vector and this iteration is complete. $P_v^{\lambda+1} = \{(u, 1/4), (v, 2/4), (x, 1/4)\}$. If the iteration $\lambda + 1$ is the last step of the algorithm, node v belongs to clusters c_u, c_v, c_x simultaneously. The mobile social clustering algorithm is detailly described in Algorithm 1.

3.2 Hierarchical Structure Detection

Hierarchical structure is an potential characteristic of industrial IoT. Based on the results of the last subsection, we propose a mobile hierarchical structure detection algorithm. Firstly, nodes that belong and only belong to a cluster will be removed from industrial IoT. The rest of nodes in the cluster are given a same label (different cluster contains different label). Secondly, randomly select one node ϕ as the listener. All neighbors of node ϕ send their labels to the node ϕ and node ϕ choose the label which has the maximum number from the received labels. If each node in the network has a label with the maximum number from their neighbors, the iteration is stopped. Otherwise, repeat this step. Finally, if the rest of nodes have the same label, the algorithm stops, if not, go to the first step. The flow diagram of this algorithm is described in Fig. 4.

When the algorithm is completed, the dendrogram will be constructed. The potential relationship in industrial IoT can be detected and all mobile devices are divided into different mobile social clusters. As mentioned above, any mobile devices can belong to more than one cluster. We can get the clear perspective for industrial IoT.

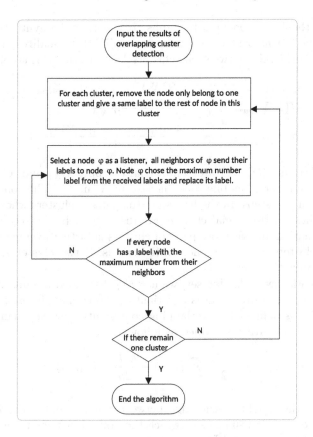

Fig. 4. Flow chart of industrial cloud hierarchical structure detecion.

3.3 Time Complexity

As industrial IoT contains millions of mobile devices, time complexity is an important specification for the mobile social cluster detection algorithm. In our algorithm, the major time consumption is focused on the overlapping mobile social cluster detection. When we extract the PCs for mobile device h, d_h comparisons are made to discover the neighbors of h with the maximum number of common neighbors. The number of comparisons will be at most $\sum_{h=1}^{n} d_h$ where d_h represents the degree of node h and n is the number of mobile devices. On the other hand, $\frac{1}{2}\sum_{h=1}^{n} d_h = m$, where m is the total of links and T is the sum of iteration. In short, the complexity of OMSC is $O(Tm)$.

4   Experiments and Results

In this section, to validate the performance of the algorithm OMSC, we conducted experiments in synthetic networks and real-world network dataset.

We adopted the LFR benchmark graphs to generate the synthetic networks dataset and the Normalized Mutual Information (NMI) to qualify our algorithms performance [16,17]. Given two social clusters A and B, the NMI is defined below.

$$
\begin{aligned}
NMI(X,Y) &= 1 - \tfrac{1}{2}(H(X|Y)_{part} + H(Y|X)_{part}) \\
H(X|Y)_{part} &= \tfrac{1}{|C_X|} \sum_i \frac{\min_{j \in \{1,2,\ldots,|C_Y|\}} H(X_i|Y_j)}{H(X_i)} \\
H(Y|X)_{part} &= \tfrac{1}{|C_Y|} \sum_i \frac{\min_{j \in \{1,2,\ldots,|C_Y|\}} H(Y_i|X_j)}{H(Y_i)}
\end{aligned}
\tag{4}
$$

where $H(X|Y)$ and $H(Y|X)$ are conditional entropy, $|C_X|$ and $|C_Y|$ severally represent the number of mobile social cluster in X and Y. The computation of NMI consists of two steps. In the first step, the pairs of clusters whose similarity degree is highest in two social clusters are discovered. In the second step, the mutual information among those pairs of clusters is further averaged. The more similar two clusters are, the higher NMI score is. If two cluster A and B are exactly the same, the value is 1.

For real-world networks dataset, we use EQ (extended modularity) to evaluate the cluster structures. EQ is a variant of the commonly used modularity (Q) [18], which is defined for overlapping communities by Newman et al. [19]. This extended modularity is defined as follows:

$$
EQ = \frac{1}{2m} \sum_{\alpha,\beta} (W_{\alpha\beta} - \frac{k_\alpha k_\beta}{2m}) \ell(c_\alpha, c_\beta)
\tag{5}
$$

where $W_{\alpha\beta}$ is the weight of the edge between node α and node β. k_α is the degree of vertex α. c_α is the cluster which the node α belongs to. m is the sum of the edges of the industrial IoT,

$$
m = \frac{1}{2} \sum_{\alpha,\beta} W_{\alpha\beta}
\tag{6}
$$

and the function $\ell(c_\alpha, c_\beta)$ is 1 iff $c_\alpha = c_\beta$, that is, α and β belong to the same cluster and 0 means other situations.

4.1   Synthetic Networks Dataset

As shows in Fig. 5, we tested OMSC algorithm in the networks which were generated by LFR benchmark. N is the number of nodes in the synthetic network. γ is the exponent for the cluster degree distribution, ϑ is the exponent for the cluster size distribution, and $\langle k \rangle$ is the average degree and there is a mixing parameter μ, which is the average fraction of neighboring mobile devices of a mobile device that do not belong to any cluster that benchmark mobile device belongs to. This parameter controls the fraction of edges that are between clusters. As shown in Table 1, we choose the $N = 4000$ and 8000, $\gamma = 2$ and $\gamma = 3$, $\vartheta = 1$ and $\vartheta = 2$ and μ from 0.1 to 0.5 respectively. The result showed that OMSC achieves the most stable and competitive NMI scores in synthetic industrial IoT.

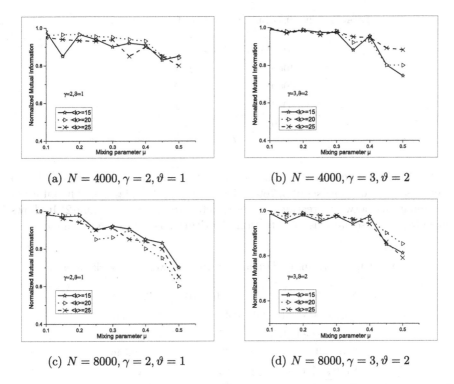

(a) $N = 4000, \gamma = 2, \vartheta = 1$ (b) $N = 4000, \gamma = 3, \vartheta = 2$

(c) $N = 8000, \gamma = 2, \vartheta = 1$ (d) $N = 8000, \gamma = 3, \vartheta = 2$

Fig. 5. Test of our algorithm on the LFR benchmark. The number of nodes is N, γ is the exponent for the degree distribution, ϑ is the exponent for the cluster size distribution, and $\langle k \rangle$ is the average degree.

Moreover, with the increasing of mixing parameter μ, the NMI scores still remains high and balance. we can draw a conclusion that the clusters produced through OMSC maintain the pretty high similarity with the ground-truth, even though many social clusters are overlapped with each other.

4.2   Real-world Networks Dataset

We next utilize OMSC to make an analysis of the real-network datasets which are more irregular than synthetic networks and find their overlapping mobile social clusters. Because industrial IoT and social networks have the same characteristics and there are no real industrial IoT datasets available, we use the four real social network datasets, namely the Zachary's karate club network, the bottlenose dolphin network, the American college football league and the network of users of the Pretty-Good-Privacy algorithm, to verify our algorithm. Among them, the well-known Zachary's karate club network contains 78 edges and 34 nodes [20]. There are 159 edges and 62 nodes in the bottlenose dolphin network [21]. The American college football league is composed of 613 edges and 115 nodes [22]. The network that implements the Pretty-Good-Privacy algorithm

Table 1. The distribution number of nodes γ and ϑ.

Number of Nodes	γ	ϑ
4000	2	1
4000	3	2
8000	2	1
8000	3	2

has 24340 edges and 10680 nodes. Besides, the communicating network based on emails in Enron consists of 367662 edges and 367692 nodes [23]. The information about the networks testing results is detailedly shown in Table 2. It shows that our mobile social cluster detection algorithm has a good performance.

Table 2. Our algorithm performance in the real-world network datasets.

Network	Modularity(Q)	Time (s)
Zacharys karate club	0.45	0.392
Bottlenose dolphin	0.57	0.481
Football	0.63	0.546
Pretty-Good-Privacy	0.82	125.45
Enron via emails	0.60	3256.26

5   Conclusion

In this paper, we propose a mobile overlapping social cluster and hierarchical structure detection algorithm in industrial IoT, which can discover the mobile social clusters with near-line time and the results of this algorithm give a new perspective for Industrial IoT. We apply this algorithm to synthetic network and real-world network dataset, and the experiment result shows that our algorithm has a high performance. The algorithm promotes the development of industrial IoT and offers great help for researchers to intensively study industrial IoT. In future work, we plan to further increase the performance of our algorithm which can be used in the world-class industrial IoT.

Acknowledgments. This work was supported by the National Natural Science Foundation of China under Grant No. 61103234 and No. 61272417, the China Scholarship Council and the Fundamental Research Funds for the Central Universities under Grant No. DUT16QY18.

References

1. Liu, J., Li, Y., Chen, M., Dong, W., Jin, D.: Software-defined internet of things for smart urban sensing. IEEE Commun. **53**(9), 55–63 (2015)
2. Wang, S., Wan, J., Li, D., Zhang, C.: Implementing smart factory of Industrie 4.0: an outlook. Int. J. Distrib. Sens. Netw. **2016**, 10 (2016)
3. Al-Saeed, L., Li, M., Al-Raweshidy, H.: Cognitive data routing in heterogeneous mobile cloud networks. In: 2014 2nd IEEE International Conference on Mobile Cloud Computing, Services, and Engineering (MobileCloud), Washington, DC, USA, pp. 194–199 (2014)
4. Xu, C., Gong, X., Yang, L., J. Zhang.: A social group utility maximization framework with applications in database assisted spectrum access. In: IEEE INFOCOM 2014 Proceedings, Toronto, pp. 1959–1967 (2014)
5. Newman, E.J., Girvan, M.: Finding and evaluating community structure in networks. Phys. Rev. E **69**(2), 292–313 (2004)
6. Fortunato, S.: Community detection in graphs. Phys. Rep. **486**(35), 75–174 (2009)
7. Huang, J., Sun, H., Han, J., Feng, B.: Density-based shrinkage for revealing hierarchical and overlapping community structure in networks. Phys. A Stat. Mech. Appl. **390**(11), 2106–2171 (2011)
8. Souam, F., Aïtelhadj, A., Baba-Ali, R.: Dual modularity optimization for detecting overlapping communities in bipartite networks. Knowl. Inf. Syst. **40**(2), 455–488 (2014)
9. Sun, P., Gao, L., Han, S.: Identification of overlapping and non-overlapping community structure by fuzzy clustering in complex networks. Inf. Sci. **181**(6), 1060–1071 (2011)
10. Yang, J., Leskovec, J.: Overlapping community detection at scale: a nonnegative matrix factorization approach. In: Proceedings of the Sixth ACM International Conference on Web Search and Data Mining, New York, NY, USA, pp. 587–596 (2013)
11. Psorakis, I., Roverts, S., Ebden, M., Sheldon, B.: Overlapping community detection using Bayesian non-negative matrix factorization. Phys. Rev. E **83**(6), 1–9 (2011)
12. Zhang, Y., Yeung, D.: Overlapping community detection via bounded nonnegative matrix tri-factorization. In: Proceedings of the 18th ACM SIGKDD International Conference on Knowledge Discovery and Data Mining, New York, NY, USA, pp. 606–614 (2012)
13. Raghavan, U.N., Albert, R., Kumara, S.: Near linear time algorithm to detect community structures in large-scale networks. Phys. Rev. E **76**(3), 1–11 (2007)
14. Steve, G.: Finding overlapping communities in networks by label propagation. New J. Phys. **12**(10), 1–27 (2010)
15. Zhang, Z., Wang, Z.: Mining overlapping and hierarchical communities in complex networks. Phys. A Stat. Mech. Appl. **421**, 25–33 (2015)
16. Lancichinetti, A., Fortunato, S., Radicchi, F.: Benchmark graphs for testing community detection algorithms. Phys. Rev. E Stat. Nonlin. Soft Matter Phys. **78**(4), 561–570 (2008)
17. Estévez, P., Tesmer, M., Perez, C., Zurada, J.M.: Normalized mutual information feature selection. IEEE Trans. Neural Netw. **20**(2), 189–201 (2009)
18. Blondel, V.D., Guillaume, J., Lambiotte, R., Lefebvre, E.: Fast unfolding of communities in large networks. J. Stat. Mech. Theor. Exp. **2008**(10), 1–12 (2008)
19. Newman, M.E.: Analysis of weighted networks. Phys. Rev. E Stat. Nonlin. Soft Matter Phys. **70**(5), 1–9 (2004)

20. Fortunato, S.: Community detection in graphs. Phys. Rep. **486**(3), 75–174 (2010)
21. Psorakis, I., Roberts, S., Ebden, M., Sheldon, B.: Overlapping community detection using Bayesian non-negative matrix factorization. Phys. Rev. E **83**(6), 1–9 (2011)
22. Xie, J., Kelley, S., Szymanski, B.K.: Overlapping community detection in networks: the state-of-the-art and comparative study. ACM Comput. Surv. (CSUR) **45**(4), 1–35 (2013)
23. Leskovec, J., Lang, K.J., Dasgupta, A., Mahoney, M.W.: Community structure in large networks: natural cluster sizes and the absence of large well-defined clusters. Internet Math. **6**(1), 29–123 (2009)

Junction Based Table Detection in Mobile Captured Golf Scorecard Images

Junying Yuan[✉], Haishan Chen, Huiru Cao, and Zhonghua Guo

Department of Electronic Communication and Software Engineering,
Nanfang College of Sun Yat-Sen University, Guangzhou 510975, China
cihisa@outlook.com
http://www.scholat.com/cihisa

Abstract. Table detection in mobile captured images faces many challenges owning to the well-known low image quality. Recently, a few researches pioneer in detecting the tables in rich-text images, but few works for scorecard images which usually lack of texts but are rich in graphics, such as golf scorecard images. In this paper, a junction-relation based table detection method for mobile captured scorecard images is proposed. Firstly, the most distinguished junctions are determined via a simplified pattern matching method, then the fault detections are removed through filtering operations, finally the missed junctions are recovered utilizing the pair-wise relationships among neighboring junctions. The experimental results show that 98.47 % of the junctions from 90 test images are correctly detected, and thus proves the superiority of the proposed method.

Keywords: Mobile captured images · Junction detection · Table detection · Pair-wise relationship · Junction filtering · Junction recovery

1 Introduction

Table is one of the most commonly used document types due to the simple but well-defined structure for highly concise data representation [1]. Therefore, automatic table recognition has been extensively studied in the past years and are broadly applied in the business and the daily life. Traditional table image analysis methods [2–5] focus on flatbed scanned images, which are acquired with high quality and thus are easy to recognize. They usually assume that straight and parallel lines exist in the scanned images, e.g. the table ruling lines and the text lines. Based on this assumption, tables are usually recognized satisfactorily.

In recent years, smart phones with high performance digital cameras become widely used in every corner of the society. Compared to flatbed scanners, smart phones are portable, fast in response and rich in functionality [6,7]. As a result, requirements arise in demanding for realtime document image analysis using smart mobiles. However, the quality of mobile captured (MC) document images

© ICST Institute for Computer Sciences, Social Informatics and Telecommunications Engineering 2016
J. Wan et al. (Eds.): Industrial IoT 2016, LNICST 173, pp. 179–188, 2016.
DOI: 10.1007/978-3-319-44350-8_18

are considerably low due to known issues, such as low effective resolution, perspective distortion, shading and noise. Therefore, traditional table analysis methods are no longer effective when applying them directly to mobile captured table images [1,8]. Seo et al. [8] proposed a junction-based table detection method for camera captured images. This method firstly locates the table region using a modified X-Y cutting technique, then detects the intersections of horizontal and vertical ruling lines, and finally label the junctions using a cost function which is constructed based on the directional information. Seo et al.'s method works well for document images with simple backgrounds, where no complex table content or text touching exists.

The contents of golf scorecards vary in table format, background and font formats. In addition, the quality of the mobile captured pictures are considerably low, such that the ruling lines are often blurred and texts sometimes touch the ruling lines. Yuan et al. [9] made a first attempt in table region detection and ruling line detection in mobile captured golf scorecard images. However, the method produces a number of miss detections and fault detections. This manuscript focuses on the problem of table detection in mobile captured golf scorecard images. Firstly, the score table region is detected by extracting the maximum connected component (CC) in the binary image [9]. Then, the candidate junctions (those pixels which have partial directional information) are captured using a simplified pattern matching technique, and are further filtered to peak out the true detections by exploiting the pair-wise junction relationships. Finally, the pair-wise relationships among neighboring junctions are further exploited to recover the missed detections. Experimental results show that the proposed method can detect **98 %** of the real junctions while generating few fault detections.

The rest of this manuscript is organized as follows. Section 2 briefs the related works and the observations from mobile captured golf scorecard images. The proposed work is detailed in Sect. 3. The experimental results are analyzed in Sect. 4. Section 5 concludes this manuscript.

2   Related Works and Observations

This section firstly briefs Seo et al.'s work in [8] and the previous work in [9], and then lists the observations from mobile captured golf scorecard images.

2.1   Related Works

Seo et al. consider junctions as the intersections of horizontal and vertical ruling lines, and label them into 12 patterns based on the direction information. To improve detection performance, they constructed a cost function based on the inter-junction connectivity information, and then minimized it using belief propagation algorithm. Their method is verified on a newly built image set which is composed camera captured magazines and books. The tables therein is simple in table structure, content, text font and background. In addition, there is no text touching issues in the binary table images.

The research in [9] handles with junction detection in mobile captured golf scorecard images, where the problems of diverse thickness table ruling lines, complex fonts and backgrounds, exist commonly. The table region is easily detected by finding the maximum connected components in the binary image, and the binary ruling lines are enhanced before junction detection which is performed using a simplified pattern matching technique. However, problem still remains in the commonly existed fault detections and miss detections due to text touching and the poor binarization. For further details of the above two methods, please refer to [8] and [9] respectively.

2.2 Observations

In this section, distinguishable observations from the mobile captured golf scorecard images are listed. Some of them may block junctions from being correctly detection, and the others could on aid the process of junction detection. As illustrated in Fig. 1, the problems blocking junction detection are listed as below:

- Table contents vary vastly in table format, fonts and background.
- Perspective distortion, shading, noise and motion blurs exist commonly in the mobile captured score table images.
- The effective resolution is low.
- Text touching exists in a part of the mobile captured images.

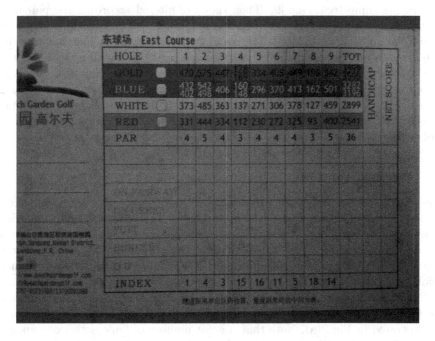

Fig. 1. A typical mobile captured golf scorecard image

In addition, some beneficial observations are also found and detailed as below:

- The score table ruling lines make up the maximum connected component in the binary image.
- The ruling lines, especially the vertical ruling lines, can be approximated as locally straight.
- Scorecards are usually taken with a very small rotation angle and thus can be neglected.

Most of the observed items have been handled or exploited in [9]. However, the problem of text touching, which is the cause of most fault detections, and the issue of miss detections, are not resolved. In this paper, the beneficial observations are further exploited to simplify the process of junction detection, the previously observed issues are carefully handled to improve the performance, and the text touching issue is resolved to minimize fault detections.

3   The Proposed Work

There remains two challenges, including the poorly binarized image and text touching, before moving table recognition in mobile captured scorecard images to the real world. The problem of poor binarization may reduce miss detections and fault detections, and text touching always results in numerous fault detections.

In this manuscript, candidate junctions are firstly detected via the simplified pattern matching technique [9]. Then they are filtered according the pair-wise relationships where junctions are adjacent and are horizontally or vertically connected via ruling lines. In order to avoid fault detections, a junction filtering step removes as many candidate junctions as possible, including some of the true detections. Therefore the vast majority of remaining junctions are true detections. Finally, the missed junctions are recovered iteratively by exploiting the existence probability of the connecting ruling lines. Accordingly, the junctions can be easily labeled using the pairwise relationships before extracting the table cells.

3.1   Junction Detection

The resolution of mobile captured images is usually high, but is often low in resolution effectiveness due to the existence of noise. Therefore, the high resolution mobile captured table images are first sampled down and then de-noised using a low-pass Gaussian filter. The table region is extracted from the maximum connected component in the binary image, and the binary ruling lines are thinned before junction detection.

The candidate junctions are captured using the simplified pattern matching technique [9] which produces many fault detections and miss detections, as illustrated in Fig. 2(a). Note that a great number of fault detections appear around the true junctions. Therefore, the candidate junctions can be clustered to select the most probable one by exploiting the directional information in the four directions, including N(orth), (W)est, (S)outh and (E)ast. The directional information can be calculated as the existing probability of a ruling line in a given direction $D \in \{N, W, S, E\}$. Let I denotes the binary table region, L_R denotes the region size, i_0 be the binary pixel where the candidate junction locates, and i_l^D be the binary pixel in the direction D of i_0 with an offset $l \in \{1, ..., L_R\}$. The directional information p_D in direction D is calculated using the equation

$$p_D^i = \frac{1}{L_1} \sum_l i_l^D. \tag{1}$$

According to the directional information, the probability of pixel i being a true detection can be determined using the formula

$$p^i = \frac{1}{4} \sum_D p_D^i. \tag{2}$$

As a result, the most probable true detections can be peaked out by finding the candidate junctions with the maximum p^i in a local region. The clustering process is executed recursively until that the inter-junction distance between any two junctions is larger than L_1, as illustrated in Fig. 2(b). The clustered junction list is denoted as J_1 for future usage.

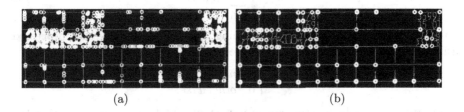

(a)   (b)

Fig. 2. Junction detection and clustering

3.2   Junction Filtering

There may still exist fault detections in junction list J_1 due to the existence of text touching and blurs, as illustrated in Fig. 2(b). Fortunately, they can be filtered out by exploiting the pair-wise junction relationships.

According to the observations, table junctions are the intersections of the horizontal and vertical ruling lines. And thus they can be arranged in a two dimensional (2-D) grid. Non-null grid element indicate true detections, and a null element reveals possibly a miss detection or the existence of no junction. Therefore, the problem of junction filtering is turned to mapping the junctions in J_1 into a 2-D grid G. Since the image rotation is negligible and the ruling lines are less distorted in the vertical direction, this mapping begins by arranging junctions column-wisely from the left to the right. And then follows the determination of horizontal pair-wise relationships.

Junction Filtering in the Vertical Direction. Before arranging junctions column-wisely, the lateral ordinates of vertical ruling lines should be determined. Since the ruling lines are no longer straight, their lateral ordinates cannot be exactly calculated. Therefore, in this manuscript, it is proposed to estimate them by exploiting the junction distribution density from junction projection in the vertical direction. Firstly, the junction density f_d is obtained by projecting the junctions in the vertical direction, then the number of junctions on a vertical ruling line is estimated, by summing up the junction density within a small range of horizontal ordinates, which is restricted by the width of the vertical ruling line. Let x_{TL}, x_{BL}, x_{TR} and x_{BR} denote the lateral ordinates at the top left corner, the bottom left corner, the top right corner and the bottom right corner, respectively. The maximum deviation $\dot{R}_x$ of the vertical ruling lines in the horizontal direction can be estimated using the equation

$$\dot{R}_h = \frac{1}{2}max\{abs(x_{TL} - x_{BL}), abs(x_{TL} - x_{BL})\}. \tag{3}$$

Since $\dot{R}_h$ indicates the maximum distance that a junction deviates from the corresponding vertical ruling line, the number of junctions on the vertical ruling line can be counted using the formula

$$n_J = \sum_{x-\dot{R}_h}^{x+\dot{R}_h} f_d(x). \tag{4}$$

To alleviates the impacts of text touching and noise, n_J is further processed using a high-pass filter specified by the equation

$$n'_J = \begin{cases} n_J & n_J > \tau_1, \\ 0 & otherwise, \end{cases} \tag{5}$$

where τ_1 is threshold to remove the impact of text touching and noise in the regions with no vertical ruling lines.

By far, the lateral ordinates of the vertical ruling lines can be estimated by the peaks of n'_J, and are denoted by $x_v(c), c \in \{1, 2, ..., N^V\}$, where N^V is the number of estimated vertical ruling lines. Given the lateral ordinates, the junction list J_1 can be filtered to remove possible fault detections, whose lateral ordinates fall out of the scopes $[x_v(c) - \dot{R}_h, x_v(c) + \dot{R}_h]$, and the remaining junctions in the scope of $[x_v(c) - \dot{R}_h, x_v(c) + \dot{R}_h]$ is called a junction column $J(c)$, where $1 \leq c \leq N^V$.

The updated junction list still contains fault detections and can be further filtered by exploiting the pair-wise relationship in the vertical direction. Since the vertical ruling lines are less distorted, the neighboring junctions on the same vertical ruling line presents a very close slope factor. Let $\phi(j, k)$ represent the slope factor between the j-th and k-th junctions in the same column $J(c)$, the standard deviation σ_ϕ of the pair-wise junction slope factors can be calculated using the equation

$$\sigma_\phi = \frac{\sum_j \sum_{k \neq j} (\phi(j, k) - \mu_\phi)^2}{N_c(N_c - 1)}, \tag{6}$$

where N_c is the number of junctions in that vertical column $J(c)$, and μ_ϕ is the average slope factor of the junction column. Similarly, the standard deviation $\sigma_\phi(j)$ of the slope factors between the j-th junction and the remaining junctions in $J(c)$, can be calculated using the formula

$$\sigma_\phi(j) = \frac{1}{N_c - 1} \sum_{k \neq j} (\phi(j,k) - \mu_\phi(j))^2, \tag{7}$$

where

$$\mu_\phi(j) = \frac{1}{N_c - 1} \sum_{k \neq j} \phi(j,k). \tag{8}$$

If $\sigma_\phi(j) < \sigma_\phi$, the j-th junction is possibly a fault detection and is removed from junction column $J(c)$. After vertical junction filtering, the updated junction list is denoted as J_2.

Junction Filtering in the Horizontal Direction. Since ruling lines can be treated as locally straight, the pair-wise junction relationships in the horizontal direction, can be determined by the existence probability of a horizontal ruling line. Let ζ_c^r denotes a junction in junction column $J(c)$ with index r, x_r and y_r be the lateral and vertical ordinates of junction ζ_c^r. The existence probability of a horizontal ruling line between ζ_c^r and $\zeta_{c+1}^{r'}$, can be calculated using the formula

$$
\begin{aligned}
y &= y_r + \left\lfloor \frac{(y_{r'} - y_r)}{(x_{r'} - x_r)} * (x - x_r) \right\rfloor, \\
p_h(\zeta_c^r, \zeta_{c+1}^{r'}) &= \frac{1}{x_{r'} - x_r} \sum_{x=x_r}^{x_{r'}} i(x, y),
\end{aligned}
\tag{9}
$$

where $i(x, y)$ is the binary pixel value at ordinates (x, y). If $p_h(\zeta_c^r, \zeta_{c+1}^{r'})$ exceeds a certain threshold τ_2, a horizontal ruling line possibly exists between these two junctions. And if there exists a $\dot{p}_h(\zeta_c^r, \zeta_{c+1}^{r'})$ meeting the below rule, the junction $\zeta_{c+1}^{r'}$ is deemed as horizontally linked to ζ_c^r:

$$\dot{p}_h(\zeta_c^r, \zeta_{c+1}^{r'}) = max\{p_h(\zeta_c^r, \zeta_{c+1}^j, j \in [1, N_{c+1}]\} > \tau_2. \tag{10}$$

When all horizontal pair-wise junction relationships are determined, the junction grid G is constructed, as illustrated in Fig. 3, where it can be noted that all fault detections (in Fig. 2(b)) have been filtered.

3.3 Junction Recovery

The junction filtering process removes a vast majority of the fault detections, and also drops some true detections. Therefore, it is required to recover the missed detections in the junction grid G. Fortunately, this is achievable by exploiting the pair-wise relationships between neighboring junctions. It can be observed that the true detections have at least one neighboring junction in the horizontal

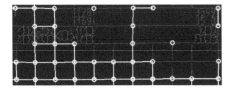

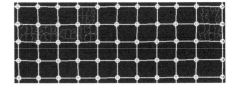

Fig. 3. Constructed junction grid   **Fig. 4.** Recovered junction grid

direction and another one in the vertical direction. Let $\zeta(r, c)$ be an element of junction grid G in the r-th row and c-th column. $\zeta(r, c)$ can be deemed as a true detection if any one of the following conditions is satisfied:

$$(p_N(r + 1, c) \geq \tau 3) \times (p_W(r, c - 1) \geq \tau 3) = 1.$$
$$(p_N(r + 1, c) \geq \tau 3) \times (p_E(r, c + 1) \geq \tau 3) = 1.$$
$$(p_S(r - 1, c) \geq \tau 3) \times (p_W(r, c - 1) \geq \tau 3) = 1.$$
$$(p_S(r - 1, c) \geq \tau 3) \times (p_E(r, c + 1) \geq \tau 3) = 1.$$

The ordinates of a recovered junction can be estimated via the neighboring junctions. The recovery process is performed repeatedly until no new junction is recovered, and the recovered and labeled junctions of Fig. 3 is shown in Fig. 4.

4   Experiments and Analysis

A few experiments are designed to verify the performance of the proposed work. The image database in [9] is employed in this manuscript. This database consists of 90 mobile captured golf scorecard images, which are captured from three different brands of mobile phones. There are 90 score tables, 1203 rows, 1188 columns and 15288 junctions in the captured images. Therein, the original and the full picture of Figs. 2, 3 and 4 is illustrated in Fig. 1, from which the issues in a mobile captured scorecard image can be easily observed.

The specific values of the three thresholds, which are employed in this manuscript, are fixed as $\tau_1 = 5$, $\tau_2 = 0.8$ and $\tau_3 = 0.25$ using empirical values from experiments. To evaluate the performance of junction detection, three measures are employed, including the precision P, the recall rate R, and the F-measure F which is the combination of P and R. The three measures are defined using the equation

$$P = \frac{tp}{N_D}, R = \frac{tp}{N_R}, F = \frac{2 \times P \times R}{P + R}, \tag{11}$$

where tp is the number of true detections, N_D is the number of detected junctions including true detection and fault detection, and N_R is the real number of junctions. The F-measure, which is the mixture of P and R, is susceptible to a minimum value, therefore it can better measure the overall performance.

Table 1 gives the performance when applying the three measures to the detected tables, rows and columns. Note that all the table regions are correctly detected. When excluding the short ruling lines in the split table cells, 99.00 % percent of the rows and 99.83 % percent of the columns can be captured successfully. Such a result reveals that, compared to that in [9], the proposed method may probably induce less miss detections.

Table 1. Detected table regions, rows and columns

	P (%)	R (%)	F (%)
Table	100.00	100.00	100.00
Row	99.00	100.00	99.50
Column	99.83	99.92	99.87

To compare with [9] (denote it as Method I), the three measures are applied to evaluate the performance when detecting all junctions J_{all}, the junctions on the external boundary J_{eb} and junction within the table external boundary J_{in}. Table 2 details the produced performance metric. Note that the proposed method overpasses Method I in each and every of the measure results. The precision of J_{all} of the proposed method is 98.47 %, which is 2.19 % higher than that of Method I. This means that much more junctions are correctly detected by the proposed work. The recall of J_{all} is 98.91 % and is 7.43 % higher than that of Method I, meaning that only a few fault detections are produced. In addition, the F-measure of J_{all} is 98.69 %, indicating the robustness of the proposed method. In a summary, the proposed method is effective in junction detection by filtering out the fault detections and exploiting the true detections.

Table 2. Detected junctions

		P (%)	R (%)	F (%)
J_{all}	Method I	96.38	91.48	93.39
	Proposed	98.47	98.91	98.69
J_{eb}	Method I	92.74	99.69	95.67
	Proposed	98.42	98.83	98.62
J_{in}	Method I	95.26	92.19	92.91
	Proposed	98.49	98.95	98.72

5   Conclusion

This manuscript presents a junction detection scheme for mobile captured scorecard images. The proposed method handles with two problems including fault detections and miss detections. By exploiting the pair-wise relationships among neighboring junctions, the vast majority of fault detections are firstly filtered out during junction grid construction, and the missed detections are then repeatedly recovered. The experimental results show that 98.47 % of the junctions can be correctly detected, and thus the effectiveness of the proposed work is proved.

Acknowledgements. This work is supported by the Foundation for Distinguished Young Talents in Higher Education, the Teaching Quality and Teaching Reform Project the Science and Technology Project of Guangdong, China, with Grant Nos. 2013LYM0123, ZL2013025, and 2013B090500067 respectively. Any options, findings, and conclusions or recommendations expressed in this paper are those of the authors. Also the authors would like to thank Guangzhou Gaoyou-Box Ltd. for the support on the golf scorecards.

References

1. Kise, K.: Handbook of Document Image Processing and Recognition. Springer, London (2014)
2. Chen, J., Lopresti, D.: Ruling-based table analysis for noisy handwritten documents. In: Proceedings of the 4th International Workshop on Multilingual OCR. ACM (2013)
3. Gatos, B., Pratikakis, I., Perantonis, S.J.: An adaptive binarization technique for low quality historical documents. In: Marinai, S., Dengel, A.R. (eds.) DAS 2004. LNCS, vol. 3163, pp. 102–113. Springer, Heidelberg (2014)
4. Cesarini, F., Marinai, S., Sarti, L., Soda, G.: Trainable table location in document images. In: Proceedings of the 16th International Conference on Pattern Recognition, vol. 3, pp. 236–240. IEEE (2002)
5. Ha, J., Haralick, R.M., Phillips, I.T.: Recursive X-Y cut using bounding boxes of connected components. In: Proceedings of the Third International Conference on Document Analysis and Recognition, vol. 2, pp. 952–955. IEEE (1995)
6. Liang, J., Doermann, D., Li, H.: Camera-based analysis of text and documents: a survey. Int. J. Doc. Anal. Recogn. (IJDAR) **7**(2–3), 84–104 (2005)
7. Mirmehdi, M.: Special issue on camera-based text and document recognition. Int. J. Doc. Anal. Recogn. (IJDAR) **7**(2–3), 83 (2005)
8. Seo, W., Koo, H., Cho, N.: Junction-based table detection in camera-captured document images. Int. J. Doc. Anal. Recogn. (IJDAR) **18**(1), 47–57 (2015)
9. Yuan, J.Y., Chen, H.S., Cao, H.R.: An efficient junction detection approach for mobile-captured golf scorecard images. Procedia Comput. Sci. **55**, 792–801 (2015)

Developing Visual Cryptography for Authentication on Smartphones

Ching-Nung Yang[1(✉)], Jung-Kuo Liao[1], Fu-Heng Wu[1], and Yasushi Yamaguchi[2]

[1] National Dong Hwa University, Hualien, Taiwan
cnyang@mail.ndhu.edu.tw
[2] University of Tokyo, Tokyo, Japan
yama@graco.c.u-tokyo.ac.jp

Abstract. Visual cryptography scheme (VCS) is a kind of cryptography that can be directly decoded by human visual system when transparent films are stacked. It requires no computation for decryption and can be stored in physical materials such as films. Therefore VCS can be a basis for providing secure and dependable authentication scheme, because it cannot be harmed by electronic and/or computational tricks. In this paper, we develop VCS for authentication on smartphones. Several authentication schemes using VCS are designed. Image quality of VCS is an inevitable issue because of small display areas of mobile devices. Thus, we will deal with VCS for continuous-tone images (gray-scale images and color images) that can enhance the image quality of VCS, so that feasible authentication schemes for smartphones will be achieved by using this continuous-tone VCS. Our authentication scheme can avoid the inconvenience of using password everywhere in modern digital life, and also resists attacks from hackers and the man-in-middle attack. This type of authentication using VCS may have a huge impact on future authentication schemes for mobile devices.

Keywords: Visual cryptography scheme (VCS) · Continuous-tone VCS · Authentication · 2D barcode · Smartphone · Threshold scheme

1 Introduction

The visual cryptographic scheme (VCS) has been firstly proposed by Naor and Shamir [1]. VCS is often implemented as a threshold (k, n) scheme, which a secret image is subdivided into n shadow images (called shadows). Any k shadows can be simply superimposed together to recover the secret image. However, $(k\text{-}1)$ or fewer shadows cannot obtain any secret information. The attractiveness of VCS is that the reconstruction does not require any computation, and can be visually decoded via the human visual system by directly stacking shadows. This novel stacking-to-see property of VCS can be applied on securely and cheaply sharing short messages, e.g., passwords or safe-combination, in the situations where we want to recover the key without computer for some secure reasons. Although VCS cannot recover the original image without distortion, the simplicity of VCS actually provides new applications in visual authentication,

© ICST Institute for Computer Sciences, Social Informatics and Telecommunications Engineering 2016
J. Wan et al. (Eds.): Industrial IoT 2016, LNICST 173, pp. 189–200, 2016.
DOI: 10.1007/978-3-319-44350-8_19

steganography, and image encryption. For example, some VCS applications combining watermark, fingerprint, Google street view, and bar code were introduced in [2–5].

Especially, the stacking-to-see property makes VCS ideally suited for use in visual authentication, and can let a user perform verification personally. This type of authentication involving human factor actually enhances the system security like seeing-is-believing. The first visual authentication using VCS was proposed by Naor and Pinkas [6]. RcCune et al. also adopted VCS to enhance the security in logging to a wireless AP [7]. Some security criteria of VSS-based authentication are formally discussed in [8]. To enhance the recognition of PIN code in visual authentication, the segment-based VCS was introduced [9]. Other VCS-based authentication schemes can are accordingly proposed [10–12].

In this paper, we adopt the high image quality of continuous-tone VCS to print shadow on a film, which can be stuck on the smartphone screen. Finally, a simple, efficient and secure VCS-based authentication on smartphone can be achieved. Several VCS-based authentication schemes are designed. The rest of this paper is organized as follows. In Sect. 2, the notion of VCS is briefly reviewed. In Sect. 3, we propose two authentication schemes (QR-code-based authentication and content-based CAPTCHA-like visual authentication). By integrating various VCS, we design a multi-server system authentication in Sect. 4. Comparison and discussion are included in Sect. 5. The conclusion is in Sect. 6.

2 Preliminary

The main technology used in this paper is VCS, which is a kind of cryptography that can be decrypted directly with human visual system without any computation. Naor and Shamir's (k, n)-VCS [1] is implemented by $n \times m$ black and white Boolean matrices B_1 and B_0. The collection C_1 (respectively, C_0) is a set obtained by permuting the columns of B_1 (respectively, B_0) in all possible ways. When sharing a black (respectively, white) secret pixel, the dealer randomly chooses one matrix in C_1 (respectively, C_0) and select a row to a relative shadow. In a black-and-white VCS, each pixel is subdivided into m subpixels in each of n shadows. A (k, n)-VCS uses h black subpixels and $(m-h)$ white subpixels (denoted as $hB(m-h)W$), and $lB(m-l)W$, where $0 \le l < h \le m$, to represent black and white secret pixels, respectively. The values of h and l are the blackness of black color and white color.

Here, we use the $(2, 2)$-VCS with $B_1 = \begin{bmatrix} 1100 \\ 0011 \end{bmatrix}$ and $B_0 = \begin{bmatrix} 1100 \\ 1100 \end{bmatrix}$ of $m = 4$ (no aspect ratio distortion) to illustrate a simple 2-out-of-2 VCS. For each row, it is observed that we have 2B2 W, and thus shadows are noise-like. When stacking two shadows, we have 4B0 W and 3B1 W for black and white secret pixels (i.e., $h = 4$ and $l = 2$), and we can visually decode the secret image. Figure 1 illustrates this $(2, 2)$-VCS where a secret image (a chessboard-like picture in Fig. 1(a)) is reconstructed by stacking two transparent films on which random-dot-like images are printed. This kind of VCS can be realized as follows. Suppose an image on a film has pixels each of which consists of 2 by 2 subpixels where two of them are white and rests are black.

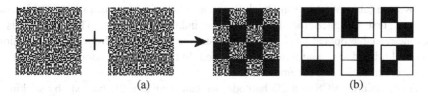

Fig. 1. A (2, 2)-VCS with $m = 4$, $h = 4$ and $l = 2$: (a) shadows and the stacked result (b) six patterns of 4-subpixl block.

There are six patterns of such subpixel arrangements as depicted in Fig. 1(b). The pattern remains the same if the same patterns are stacked. The result becomes totally black when the reverse patterns in each column of Fig. 1(b) are stacked.

3 The Proposed Visual Authentication Schemes

Our paper is not just to design VCS-based authentication schemes on smartphones. Image quality of VCS is an inevitable issue because display areas of mobile devices are very limited. Therefore, we will show how to enhance the image quality of the stacked result of VCS.

3.1 QR Code Based Authentication

Because two-dimensional (2D) barcode contains only black/white point, which is the same to the traditional VCS. So, the combination of these two technologies is reasonable. In fact, there are already some literatures on combining VCS and 2D barcode [5, 13]. In [5], the authors studied embedding the public and private information into 2D barcode by VCS. Another approach [13] adopts 2D barcode on shadows for cheating prevention. Our combination of VCS and 2D barcode is different to these two types. Our shadows are noise-like and the stacked result is 2D barcode. Table 1 shows the advantages of such combination.

Table 1. Advantages of the proposed combination of VCS and 2D barcode.

Advantage	Description
Robustness ability	The recovered image of VCS has distortion, while 2D barcode has the robustness against error. If the errors of superimposed image are within the fault-tolerant ability of 2D barcode, then barcode decoder can correctly decode the secret. Therefore, via this combination, the robustness of 2D barcode can make up for the poor image of VCS
Embedding capacity	2D barcode has the small size and meantime can carry the large and various data (such as: images, text, symbols and other types of information). This feature can make up the shortcoming of traditional VCS that only decodes a simple image
Easy decoding	2D barcodes are widely used in business, and barcode decoders are already very common, and meanwhile all smartphones can decode 2D barcode. Thus, we can easily decode 2D barcode anywhere

QR code (abbreviated from Quick Response Code) is a type of matrix 2D barcode, which is first designed for the automotive industry in Japan. This barcode is a machine-readable optical label. It has large storage capacity, high-speed identification, and small printed area, and thus is widely used. In this paper, we adopt QR code in the proposed authentication scheme.

Via combining VCS and 2D barcode, we can recover a 2D barcode by sticking a film on a smartphone screen. Then, we provide this bar code to decoder for authentication. As shown in Fig. 2(a), there are three entities in the proposed QR-code-based authentication: verification server (VS), smartphone (SP) and barcode decoder (BD). By using a (2, 2)-VCS, VS generates two shadows and keep one as a verification shadow in VS, and print the other shadow on a film. This low-cost, credit-card sized film can be stuck to smartphone screen. At this time, the film with a smartphone plays the role of a "pass token" like using the Passbook in iPhone. If the user wants to login for some sites, SP sends a request to VS. After receiving the login request, VS sends the verification shadow to SP. The SP holder put the film on smartphone screen, and submit this SP with a display of QR code to BD for authentication. Another application scenario of QR-code-based authentication is shown in Fig. 2(b). User may have various films in pocket. VS can send verification shadow to any check in counter. For various occasions, the user adopts a different film.

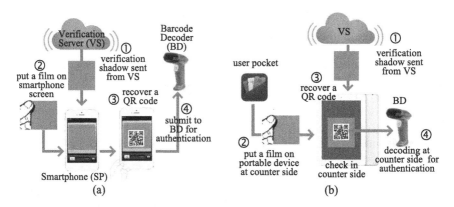

Fig. 2. QR-code-based authentication scheme (a) application scenario 1: verification shadow is sent to smartphone (b) application scenario 2: verification shadow is sent to the check in counter.

As we already mentioned, the reconstructed pixels of VCS by stacking a film are not real binary which may cause errors. If the distorted 2D barcode (the stacked result of VCS) is beyond the error-tolerant ability of QR code. The barcode decoder cannot successfully decode this barcode and the authentication fails. In this paper, we propose two solutions. The first approach is to intentionally distribute errors among a QR code according to Reed-Solomon (RS) code. Another approach is to design a new-type subpixel from the notion of continuous-tone VCS [14], to enhance the image quality of QR code. These two approaches are briefly described below.

Using Error Correcting Capability of RS Code: QR code adopts the RS code for error control coding, and it has a strong error correcting ability. There are four fault-tolerance levels of L, M, Q, H, with fault-tolerance of 7 %, 15 %, 25 % and 30 %, respectively. Obviously, the higher fault-tolerance level is, the lesser embedding data we have. Figure 3 is an example of QR code (Model 2 Version 5-H), where D blocks are the data blocks, E blocks are the parity digits for D blocks via RS code encoding. The QR code in Fig. 3 has four RS codes: two (33, 11, 11) RS codes and two (34, 12, 11) RS codes. The error correction capability are 11 for all four codes, one has the code length 33 (information length 11), and the other has the code length 34 (information length 12). For this example, QR code has fault-tolerance of 11/33 or 11/34 about 30 %.

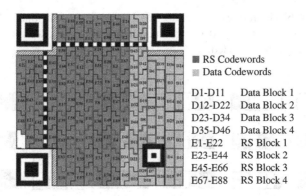

D1-D11	Data Block 1
D12-D22	Data Block 2
D23-D34	Data Block 3
D35-D46	Data Block 4
E1-E22	RS Block 1
E23-E44	RS Block 2
E45-E66	RS Block 3
E67-E88	RS Block 4

Fig. 3. Model 2 version 5-H QR Code. (Color figure online)

We study how to distribute these errors, and make errors being within the error tolerance of RS code. Here, we use a simple example (Example 1) to show the strategy of dispersing errors. As shown in Fig. 3, the red color implies the first (33, 11, 11) RS code with data blocks D1-D11 and parity blocks E1-E22. Suppose that we use perfect black VCS (PBVCS), which has the 100 % blackness for the black secret pixel but has the distorted white secret pixel. For this case, this 2D barcode can decode the correct black dot, while the decoded result of white dot may be wrong.

In Example 1, we explain how to ensure that some of the white dot in 2D barcode can be correctly decoded. However, at this time, the black dot may be not recovered. We hope that the errors can be uniformly dispersed in these four RS codes, such that the errors of each RS code does not exceed 11 (i.e., within the error tolerance), and finally the QR code is decodable.

Example 1: Consider the (2, 2)-VCS in Fig. 1. Because we have 4B for black dot, so the decoding of black dot in QR code is correct, but there might be wrong for decoding the white dot.

Figures 4(a)–(d) show all four cases of the middle secret pixel is white: (□□□), (□□■), (■□□) and (■□■). Consider the case in Fig. 4(a), we can adjust the subpixel of upper left corner in the right 4-pixel block to the upper right corner (the adjusted subpixel denoted as the gray color). This adjustment ensures the correct decoding of the middle pixel. At this time, left and right dots also has chance of decoding correctly. Figures 4(b) and (c) illustrates the adjustments of (□□■) and (■□□). For the case that the left and right are all black (■□■), we can only choose one side to change black secret pixel to white secret pixel. As shown in Fig. 4(d), the red subpixels imply changing from black color to white color. In this case, the right dot will be wrongly decoded. These adjustments are only the horizontal adjustments. To precisely disperse errors, we should consider the comprehensive adjustment with 8 around subpixels P1-P8, as shown in Fig. 4(e). In fact, we can further improve the adjustment by using

the black matrix $B_1 = \begin{bmatrix} 1100 \\ 0110 \end{bmatrix}$ with the black secret pixel 3B1 W. Because 3B1 W is

not the perfect black, it may cause decoding errors of some black secret pixels. However, at this time, Fig. 4(d) can be modified to Fig. 4(f), and finally all secret pixels may be correctly decoded.

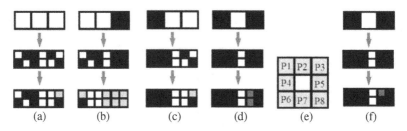

Fig. 4. Adjusting subpixels to ensure the correct decoding of the middle white secret pixel: (a) left and right are all white (□□□) (b) left white and right black (□□■) (c) left black and right white (■□□) (d) left and right are all black (■□■) (e) the comprehensive adjustment (f) the adjustment for the case (■□■) by 3B1 W. (Color figure online)

The adjustment and the choice of fault-tolerance level (Level L, M, Q, H) have great relevance. Both strategies should be completely tested and analyzed to select a best adjustment policies.

Enhancing the Image Quality of QR Code: We adopt the notion of continuous-tone VCS to design a new subpixel structure, to enhance the image quality of the QR code. Here, we use circular subpixel structure (see Fig. 5) to describe the continuous-tone VCS (note: to really implement the continuous-tone VCS, we should use the square structure [14]). Barcode decoders measure the intensity of the light reflected back from the barcode, and determines the grayness of dots. Thus, we can use the inner and outer rings, and add the gray color in Fig. 5, to adjust the grayness of the recovered pixel and ensure that each secret pixel can correctly decoded by barcode decoder.

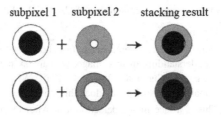

Fig. 5. The concept of the subpixel of continuous-tone VCS.

3.2 Content Based CAPTCHA-like Visual Authentication

In the QR-code-based authentication, the film with a smartphone is a "pass token" like using the Passbook in iPhone. Suppose that, for some reasons, the verifier cannot provide barcode decoder. For the case, we have to authenticate by visually decoding the content of the recovered image. Also, consider the case that adopts QR-code-based authentication in the advertising of electronic commerce by clicking. However, the proposed QR-code-based authentication cannot prevent the click fraud if an attacking software uses the shadow. We adopt the seeing-is-believing property into authentication to address problems of the non-availability of barcode decoder and the anti-bot. The proposed content-based CAPTCHA-like visual authentication is shown in Fig. 6.

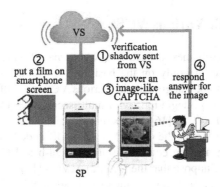

Fig. 6. Content-based CAPTCHA-like visual authentication scheme.

When using CAPTCHA for authentication, users often enter some alphanumeric words. In our content-based CAPTCHA-like authentication, the server sends natural images. Users should respond the correct answer for verification according to the reconstructed image. Because the visual authentication is content based, users have to clearly decode the image. We can use continuous-tone VCS to gain the high-quality gray/colorful image which will be integrated to the content-based authentication. The conventional CAPTCHA can include graphs, sounds, texts, pictures, smart/ mathematical query, and even videos. The most popular and simple way of CAPTCHA is the distorted texts. In fact, our content-based CAPTCHA-like visual authentication

Table 2. Query types of the proposed content-based CAPTCHA-like visual authentication.

Query type	Description
Image content identification	Consider the recovered image (a yellow rose) in Fig. 6. Based on the understanding of this image, we can ask the questions gradually, e.g., (i) What is the flower color? (ii) What is the flower? (iii) How many pieces of petals does this flower has?,..., and so on
Face recognition	This is a face image detection test. After receiving a photo from verification server, the user has to click on several face images (e.g., the same person but may be with different hair style, the sketch, or different angle of photo) to solve this face CAPTCHA
Landmark recognition	The same notion of the face recognition, but use the famous scenery all over the world
Smart/Mathematical	Based on the recovered texts, we can ask the questions by smart/mathematical CAPTCHA
Hybrid query	Combine various query types within an image

can also uses this approach. However, this will has the vulnerability that the distorted texts may be recognized by optical character recognition (OCR) software. Here, we adopt content-based challenge response. Table 2 shows some effective query types of the proposed content-based CAPTCHA-like visual authentication.

4 Integration with Various VCSs

The two authentication schemes in Sect. 3, QR-code-based scheme and content-based CAPTCHA-like scheme, are only on the basis of (2, 2)-VCS. The proposed smart-phone authentication can be integrated with various VCSs for different applications. In this section, we show two integrations with multi-secret VCS (MVCS) [15] and participant specific VCS (PSVCS) [16], respectively.

Figure 7 only demonstrates the use of MVCS in QR-code based authentication. Of course, the MVCS can also be combined into content-based CAPTCHA-like visual authentication scheme. Suppose that the (2, 2, 2)-MVCS [15] has two shadows: one is the verification shadow S_1 in server and the other shadow S_2 (printed on a film) is kept by smartphone holder. The so-called (2, 2, 2)-MVCS implies that two shadows should be stacked to recover the secret. Two secrets can be recovered from stacking S_1 and S_2, and stacking S_1 and S_2, which S_2 is the image S_2 flipped with 180°. As shown in Fig. 7, we can obtain two QR codes from $Q_1 = (S_1 + S_2)$ and $Q_2 = (S_1 + S_2)$, and can login to various databases.

Consider another application scenario, multi-server environment, which various servers can collaborate to provide different services. Here, we adopt a (t, k, n)-PSVCS [16] to design a multi-serve authentication scheme. The so-called (t, k, n)-PSVCS, where $t \leq k \leq n$, needs t specific shadows and other $(k-t)$ shadows to decode the secret image. Here we use an example of a digital content service (providing software, music, movies,..., ad etc.) to describe a multi-server system. When considering the security and the load balancing, we share the digital contents in cloud servers by secret

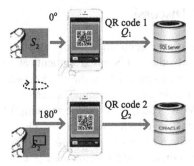

Fig. 7. Integrating (2, 2, 2)-MVCS into QR-code-based authentication for logining to various databases.

sharing and distribute them into multiple service servers (SSs). In order to protect these digital media contents and manage their digital rights, we have one or more ticket grant sever (TGS). If the user wants to login the digital content service system by his pass token (the film with his smartphone), he needs all TGSs and some SSs to get the access permission. Suppose that this digital content service system has $(t-1)$ TGSs, $(n-t)$ SSs, and need $(k-t)$ SSs agreed to provide services. The (t, k, n)-PSVCS satisfies this conditions: t specific shadows (one film kept by smartphone holder and $(t-1)$ shadows in TGS), and other $(k-t)$ common shadows $((k-t)$ SSs).

The procedure of multi-serve authentication can be implemented by either QR code based or content based. Figure 8 is an example that there are total 9 entities including one TGS (Server 1), seven SSs (Server 2–Server 8), and one user. The authentication of this multi-server system can be accomplished by the (t, k, n)-PSVCS, where $t = 2$, $k = 5$ and $n = 9$. Suppose that the shadows of Server i, $1 \leq i \leq 8$, is S_i, and that the film kept by smartphone holder is S_h. For recovering the secret information, S_1 and S_h are necessarily required and the total threshold should be five. As shown in Fig. 8, Case ① can login successfully because $\{S_1, S_2, S_3, S_4, S_h\}$ has S_1 and S_h, and the threshold is also five. However, other two cases fail. In Case ②, although $\{S_2, S_3, S_5, S_6, S_h\}$ achieves the threshold, S_1 is not involved. For Case ③, S_1 and S_h are involved but $\{S_1, S_3, S_4, S_h\}$ does not achieve the threshold.

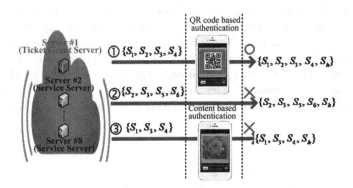

Fig. 8. Multi-server system authentication scheme.

5 Comparison and Discussion

There are some researches on authentication using VCS. A comparison between our authentication schemes and other VCS-based authentication schemes [6, 10–12] is listed in Table 1. All these schemes adopt the stacking-to-see property of VCS to achieve visual human-verifiable authentication. The schemes in [10, 11] use extended VCS (EVCS) instead of VCS. The EVCS has the meaningful cover image on shadow and give no sign that some secret data has been hidden. Our scheme and the schemes in [6, 12] have noise-like shadows. The schemes in [6, 10–12] use the simple 2-out-of-2 scheme. Our multi-server authentication scheme uses (t, k, n)-PSVCS for multi-server environment.

The major difference between our research and others is that we propose a novel technology on smartphone for authentication together with a new VCS for continuous-tone images. Our research brings the following contributions. (1) The authentication schemes can avoid the inconvenience of using password everywhere in modern digital life. The technology brings convenience and benefits for most people, because it only requires users to stack a low-cost small film onto smartphones. (2) The authentication schemes can resist any malicious attacks by crackers and the man-in-middle attacks. It is secure for multiple applications including online banking and mobile ticketing. So, the technology has a lot of business opportunities. (3) Image quality assessment for VCS can provide new criteria for image quality. The assessment is important for evaluating and controlling the image quality, because there exist trade-offs among several indices of image quality in VCS. (4) A new continuous-tone VCS may achieve better image quality than conventional VCS. It is crucial for VCS applications running on smartphones, because their display areas are very limited. (5) This technology, a low-cost, credit-card sized film stuck to smartphone, has the competitive advantages on "size" and "price". (6) The shadow image can be easily printed to various materials, and integrated into different products. So, the technology will have the large commercial value (Table 3).

As we know, there are some patents of applying VCS on smartphones. Please refer [17, 18]. Recently, the team of Liverpool Hope University announces an authentication scheme via smartphone by using VCS [17]. Also, their technology has been patented in the UK and USA and has patents pending in Canada and Europe. However, their approach is different to us. The shadow of user is sent to an APP and stored in the smartphone. Afterwards, users use this shadow to stack the received image from the server to reveal the secret code. The approach in [18] seems similar to our approach, i.e., printing shadow on a soft film. But, it does not combine 2D barcode, content based authentication, and multi-server environment. The only same idea of [18] as our approach is to print the shadow on a film. This technology [18] is now announced for sale at TYNAX website, which is a global patent and technology exchange website.

Table 3. A comparison between our authentication schemes and other VCS-based authentication schemes.

	Scheme in [6]	Scheme in [10]	Scheme in [11]	Scheme in [12]	Our scheme
VCS	(2, 2)-VCS	(2, 2)-EVCS	(2, 2)-EVCS	(2, 2)-VCS	(2, 2)-VCS (t, k, n)-PSVCS[#1]
Application scenario	Bank	Bank	Website	ID-card	Smartphone
Server/Client shadow	1/1	1/1	1/1	1/1	1/1 $(n-1)/1$[#1]
Secret image	Password	Signature	Password	ID photo	QR code[#2] Natural image[#3]
Cover image	NO	YES	YES	NO	NO
Involved servers	1	1	1	1	1 $(k-1)$[#1]
Different privilege of server	NO	NO	NO	NO	NO YES[#1]
Authentication type	Visual auth.	Visual auth.	Visual auth.	Visual auth.	Barcode auth.[#2] Visual auth.[#3]

#1 multi-server authentication scheme. #2 QR-code-based authentication scheme.
#3 content-based CAPTCHA-like visual authentication scheme.

6 Conclusion

In this paper, several authentication schemes, QR-code based authentication, content-based CAPTCHA-like visual authentication and multi-server system authentication, are designed. Meantime, we also study methods to improve image quality of VCS, on which we develop a low-cost, credit-card sized film stuck to smartphone screen. On the basis of this soft film, a simple, cheap, efficient and secure authentication can be achieved. Additionally, our technology can be combined with existing VCSs to achieve more comprehensive application scenarios.

Acknowledgments. This work was supported in part by Ministry of Science and Technology, Taiwan, under Grant 104-2918-I-259-001 and 104-2221-E-259-013.

References

1. Naor, M., Shamir, A.: Visual cryptography. In: De Santis, A. (ed.) EUROCRYPT 1994. LNCS, vol. 950, pp. 1–12. Springer, Heidelberg (1995)
2. Surekha, B., Swamy, G., Rao, K.S.: A multiple watermarking technique for images based on visual cryptography. Int. J. Comput. Appl. **1**(11), 77–81 (2010)
3. Monoth, T., Anto, P.B.: Tamperproof transmission of fingerprints using visual cryptography schemes. Procedia Comput. Sci. **2**, 143–148 (2010)
4. Weir, J., Yan, W.: Resolution variant visual cryptography for street view of google maps. In: Proceedings of ISCAS, pp. 1695–1698 (2010)
5. Yang, C.N., Chen, T.S., Ching, M.H.: Embed additional private information into two-dimensional barcodes by the visual secret sharing scheme. Integr. Comput. Aided Eng. **13**(2), 189–199 (2006)
6. Naor, M., Pinkas, B.: Visual authentication and identification. In: Kaliski Jr., B.S. (ed.) CRYPTO 1997. LNCS, vol. 1294, pp. 322–336. Springer, Heidelberg (1997)
7. McCune, J.M., Perrig, A., Reiter, M.K.: Seeing-is-believing: using camera phones for human-verifiable authentication. In: Proceeidngs of IEEE Symposium on Security Privacy, pp. 110–124 (2005)
8. Yang, C.N., Chen, T.S.: Security analysis on authentication of images using recursive visual cryptography. Cryptologia **32**(2), 131–136 (2008)
9. Borchert, B., Reinhardt, K.: Applications of visual cryptography. In: Chapter 12 of Visual Cryptography and Secret Image Sharing, Boca Raton, FL. CRC Press/Taylor and Francis (2011)
10. Jaya, Malik, S., Aggarwal, A., Sardana, A.: Novel authentication system using visual cryptography. In: Information and Communication Technologies (WICT), pp. 1181–1186 (2011)
11. Goel, M.B., Bhagat, V.B., Katankar, V.K.: Authentication framework using visual cryptography. Int. J. Res. Eng. Technol. **2**, 271–274 (2013)
12. Ratheesh, V.R., Jogesh, J., Jayamohan, M.: A visual cryptographic scheme for owner authentication using embedded shares. Indian J. Comput. Sci. Eng. **5**, 190–194 (2014)
13. Weir, J., Yan, W.: Authenticating visual cryptography shares using 2D barcodes. In: Shi, Y. Q., Kim, H.-J., Perez-Gonzalez, F. (eds.) IWDW 2011. LNCS, vol. 7128, pp. 196–210. Springer, Heidelberg (2012)
14. Nakajima, M., Yamaguchi, Y.: Enhancing registration tolerance of extended visual cryptography for natural images. J. Electron. Imaging **13**, 654–662 (2004)
15. Chen, S.K., Lin, S.J., Lin, J.C.: Flip visual cryptography (FVC) with perfect security, conditionally-optimal contrast, and no expansion. J. Vis. Commun. Image Representation **21**, 900–916 (2010)
16. Yang, C.N., Sun, L.Z., Yan, X., Kim, C.: Design a new visual cryptography for human-verifiable authentication in accessing a database. J. Real-Time Image Process. (2015). doi:10.1007/s11554-015-0511-9
17. Liverpool and Sefton Chambers of Commerce. http://www.liverpoolchamber.org.uk/article.aspx/show/5641
18. Global Patent & Technology Exchange. http://www.tynax.com/listing/4027

A Scale-Free Network Model for Wireless Sensor Networks in 3D Terrain

Aoyang Zhao, Tie Qiu[✉], Feng Xia, Chi Lin, and Diansong Luo

School of Software, Dalian University of Technology, Dalian 116620, China
qiutie@ieee.org

Abstract. The building strategy of network topology in 3D space is important technology for wireless sensor networks (WSNs). It is often more complex to assess the detection area of sensor node in 3D terrain. The research on topological modeling method, which can withstand a certain amount of node failures and maintain work properly, has become the focus in recent years. In this paper, a 3D terrain with multi-peaks is defined by Gaussian distribution and a new scale-free wireless sensor network topology is modeled in 3D terrain. According to the improved growth and preferential attachment criteria in complex network theory and considering the limited communication radius and energy of sensor nodes, this model is tolerant against random attacks and apply to WSNs in 3D terrain.

Keywords: Scale-free networks · Wireless sensor networks · 3D terrain

1 Introduction

Wireless sensor networks (WSNs) [1,2] are special networks for sensing environment and collecting information. They consist of large scale sensor nodes. Sensor nodes can collaborate monitoring, sensing and collecting environment information, which cover the geographic region, and send data to sink nodes through multi-hop. The target area of the real-world application in the WSNs is often 3-dimensional (3D) [3,4] and always has complex terrain and changeful climate, such as mountains, canyon, desert and underwater [5] etc. In recent years, with the harsh application environment and large application scale, the modeling for large-scale wireless sensor network topologies in 3D space has been brought into focus.

Complex networks [6,7] widely exist in the real world, such as global transportation networks, cooperation networks, social networks etc. As an interdisciplinary field, complex networks have attracted much attention. Small world networks [8] and scale-free networks [9,10] are classic models in complex network theory. The small world model has two characteristics for wireless network, which are smaller average path length and higher clustering coefficient. It is generally used in the modeling of heterogeneous network topologies [11]. Scale-free model is generally used in modeling homogeneous network topologies [12].

© ICST Institute for Computer Sciences, Social Informatics and Telecommunications Engineering 2016
J. Wan et al. (Eds.): Industrial IoT 2016, LNICST 173, pp. 201–210, 2016.
DOI: 10.1007/978-3-319-44350-8_20

The node degree in scale-free networks follows power-law distribution. The majority of nodes have low degree. When random attack happened, the low degree nodes have a higher probability of being attacked, the network topology will not lost too many edges. Scale-free model compared with small world model is more strongly tolerant against random attacks [13]. Thus, how to construct the scale-free network topologies in wireless sensor networks is a urgent problem, which needs to be resolved.

In this paper, we reduce the actual 3D terrain to inerratic and continuous curved surface. we consider the limitation of communication radius in WSNs, the requirement of energy balance in distributed systems [14] and the maximum degree of nodes in topology firstly. Then, we design a modeling strategy of scale-free network topology in WSNs, which based on the BA model [9] and perceived probability. The theoretical model and analytical method conform to relevant restrictions of WSNs in 3D terrain.

The rest of this paper is organized as follows. A summary of related works is presented in Sect. 2. Section 3 describes a modeling strategy of scale-free network topologies for WSNs in 3D terrain. Section 4 describes the algorithm design in detail. We present the simulations in order to show the performance of modeling algorithm in Sect. 5. Finally, Sect. 6 summarizes the main results and discusses further works of the proposed schemes.

2   Related Works

Degree is an important factor for each node. The definition of degree k_i is the number of edges connected with node i. The dynamic changes exist in majority of networks. Nodes outside continuously join these networks and nodes inside possibly fail at the same time. Many networks have demonstrated that new joined node hopes to connect with "important" node. Thus, the new connection is not completely random (i.e. the connection probability is related to the degree of each node). In order to explore the mechanism of power-law distribution, Barabasi and Albert proposed the famous scale-free network evolution model [9]. They thought a lot of existing models did not take two important criteria into account of networks:

☐ Growth: There are new nodes continuously joined, and network size is growing.
☐ Preferential attachment: The new joined node tends to connect with node which has higher degree. This phenomenon is also known as the "Matthew Effect".

The two criteria reflect power-law distribution of network connectivity in traditional scale-free networks, such as Internet, social networks, transportation networks, etc. But limited by the harsh application conditions of distributed systems, the preferential attachment property of BA model cannot be simulate directly. Hence, many researchers focus on the application of scale-free network topologies in wireless sensor network. Liu et al. [15] proposed a complex network model based on heterogeneous network used in sensor networks. This model has both small world and scale-free properties. Du et al. [16] studied on the shortest path feature of scale-free network and designed a transportation system which is more efficient. Zheng et al. [17,18] also designed two robustness scale-free network topologies, LGEM (linear growth evolution model) and AGEM (accelerated growth evolution model). They paid attention to actual situation such as increasing or removing nodes, reconstruction edges. Jian et al. [19] proposed a new scale-free model with energy-aware based on BA model, named EABA (energy-aware Barabasi-Albert). They presented the tunable coefficients to balance the connectivity and energy consumption of sensor network topology. These references take both energy balance and transmission performance into account. It will increase data overhead during the modeling process. They also not considering the influence of 3D terrain. Thus, we need to find a tradeoff solution between scale-free property and modeling overhead in 3D space.

3 Scale-Free Model for WSNs in 3D Terrain

In order to simulate the application scenarios of WSNs in real world more visually, we simplify the terrain as regular and continuous curved surface. We assume that the surface topography in 3D terrain is regular. All sensor nodes are deployed within the area of $L * W\ m^2$ and the terrain is divided into $L * W$ grid points. We adopt the multi-peaks graph which is satisfied with Gaussian distribution to simulate the application scenarios and consider the entire surface is nonlinear and continuous.

$$Height(x,y) = max_{\{n=1,2...N_{peak}\}}K * normcdf(peak(n),0,1) \tag{1}$$

$$peak(n) = \sqrt{\rho * ((x - x(n))^2 + (y - y(n))^2)^{\frac{-\eta}{2}}} \tag{2}$$

Equation (1) describes the height value of gird point (x, y) on the 3D terrain which is composed of multi-peak value functions. The max value of overlap part will be selected. N_{peak} is the number of peaks in 3D space. The function $normcdf(a, b, c)$ describes the Gaussian feature distribution, also known as Gaussian integral. It consists of $norm$ Normal distribution and cdf probability distribution. Equation (2) describes the cumulative distribution of grid point (x, y). ρ and η are peak coefficient. $(x(n), y(n))$ is the coordinate of nth peak.

We consider the height of each peak in the range of 10 to 20 and build the 3D terrain according to ρ, η, N_{peak} and different coordinate of peaks. As can be seen in Fig. 1, $(x, y, Height(x, y))$ is the coordinate of each grid point, $x \subseteq \{1, 2, \cdots, W\}$, $y \subseteq \{1, 2, \cdots, L\}$. There are $L * W$ data points in the 3D hill terrain.

Each node in wireless sensor network topology has its own communication range, so it only possesses and uses local information in the wireless sensor networks. As a result, the new joined node maybe has not sufficient neighbors to connect. In this case, preferential attachment property cannot work. In order to solve this problem, we focus on the following two criteria during the modeling process of scale-free wireless sensor network topologies.

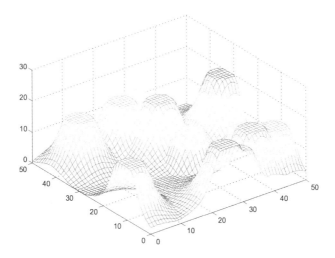

Fig. 1. The 3D terrain for scale-free networks in WSNs.

☐ Preferential attachment property is limited in the range of communication radius. Two nodes can be directly connected without the limitation of their location in Internet or other traditional scale-free networks. But in wireless sensor networks, any two sensor nodes communicate with each other through multi-hop.

☐ We control the maximum degree of nodes to prevent excessive energy consumption. In case of there are some nodes with too high degree in wireless sensor networks, they will forward and deal with a large amount of data and their energy is consumed quickly. This is a very negative phenomenon. If there is no enough energy to support, the entire wireless sensor network will soon collapse. Therefore, the maximum degree of node must be limited.

From what has been mentioned above, we consider two aspects of improvement based on the constructing criteria of BA model. Firstly, each new joined node must have sufficient neighbor nodes in its communication range. Hence, we put all the new joined nodes into the wireless sensor network topology, instead of one by one. But the nodes need asynchronously calculation during the process of adding edges. Meanwhile, neighbor nodes cannot perform the algorithm at the same time.

The perceived probability of the deployed sensor nodes is determined by the application regions of the sensor network. In Boolean perception model, the detection range of the node is in a sphere where the node is the center and the detection range is the radius. If a point falls within the radius of the sphere, the detection probability is 1, otherwise it is 0. However, the monitoring ability of each sensor node is affected by terrain, noise, signal attenuation etc. in real 3D application regions. So we use Eq. (3) to reflect the monitoring ability of nodes.

$$P_{ij} = \begin{cases} 1 \Big/ \alpha e^{\frac{\beta D_{ij}}{r}} , & (D_{ij} \leq r), \\ 0, & (D_{ij} > r). \end{cases} \tag{3}$$

P_{ij} is the perceived probability of grid point j by sensor node i. α, β are adjusted according to the physical properties of the sensor. r is the radius of the individual sensor node. D_{ij} is the Euclidean distance between sensor node i and grid point j. The bigger D_{ij} is, the lower perceived probability of grid point is. Then, we can further calculate the perceived probability P_j of each grid point j according to the number and distance of sensor nodes in Eq. (4).

$$P_j = 1 - \prod_{i=1}^{N_l}(1 - P_{ij}) \tag{4}$$

N_l is the number of sensor nodes in the perceived range of grid point j. Because the grid points are always detected by more than one sensor node and the detections of all nodes is independent, the probability of one point is accomplished by all sensor nodes that can detect it together. During the following process, only when the perceived probability of sensor node is greater than 0.9, it can be connected.

Then, we assume that all the neighbors in the communication range of a new joined node is the local-world of the new joined node. But the neighbors which already connected with the new joined node or reached the maximum degree do not belong to the local-world of the new joined node. The new joined node tends to connect with higher degree node. We define the connection probability $\prod_{Local}(i)$ for a neighbor node i in Eq. (5).

$$\prod_{Local}(i) = \frac{d_i}{\sum\limits_{i=1}^{n} d_i} \tag{5}$$

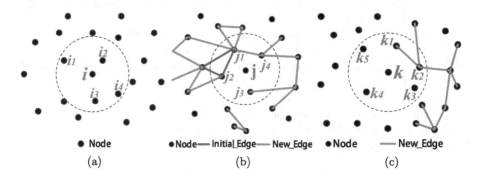

| ● Node | ●Node——Initial_Edge——New_Edge | ●Node | ——New_Edge |
| (a) | (b) | (c) |

Fig. 2. The adding connection process of new joined nodes.

Wherein, d_i is the degree of node i, and n is the total number of neighbors for the new joined node. Figure 2 describes the specific situation of connection probability during the preferential attachment process. We use the vertical projection in 2D space to explain this process more clearly. At the beginning stage, we choose 3 nodes as the initial wireless sensor network. The edges between the 3 nodes are represented by the solid red lines. The solid blue lines represent the edges, which are added during the modeling process. We assume that the perceived probability of sensor nodes in the communication range of i, j and k are all greater than 0.9. In Fig. 2a, the four neighbors in the local-world of node i are all isolated. So, when node i broadcasts the request of adding edges, it chooses m neighbors to establish connection in equal probability. However, the four neighbors in the local-world of node j have different degrees, which are 6, 3, 3 and 1 in Fig. 2b. The connection probabilities can be calculated from Eq. (5), which are 0.46154, 0.23077, 0.23077 and 0.07692. Then, it chooses m neighbors to establish connection by roulette method. In Fig. 2c, There are not only isolated nodes, but also connected nodes in the local-world of node k. The connected nodes, k_1 and k_2, whose degrees are 3 and 1, will be prior considered by node k. Node k determines the connection scheme by the value of m, whose threshold is equal to the number of connected nodes. As shown in Fig. 2c, the threshold of m is 2. If $m > 2$, in addition to establish connection with nodes k_1 and k_2, it chooses $m - 2$ neighbors to establish connection in equal probability. If $m = 2$, node k establishes connection with nodes k_1 and k_2 directly without calculation. If $m < 2$, it calculates the connection probabilities of nodes k_1 and k_2, which are 0.25 and 0.75, then uses roulette method to choose the neighbor.

4   Algorithm Design

This algorithm describes the modeling process of scale-free network topology in wireless sensor networks. The variables used in the algorithm are as follows.

☐ N_0: the number of nodes in initial wireless sensor network topology.
☐ N: the total number of nodes in scale-free wireless sensor network topology. N contains the nodes in N_0.
☐ V: the set of all nodes in scale-free wireless sensor network topology.
☐ V_0: the set of nodes in initial wireless sensor network topology.
☐ V_i: the set of nodes in the communication range of node i but not connected with it and have not reached the maximum degree.
☐ r: the communication radius of each node.
☐ P_i: the perceived probability of node i.
☐ m: the number of adding edges for each new joined node.
☐ Lst_i: list of the neighbor nodes, which connected with node i.

```
1   program BANetworkbuild (Lst_i)
2       var
3           N, V, r, P_i, m
4       begin
5       V_in = initial N_0 nodes();
6       repeat
7           broadcastStartPacket(v_i);
```

```
8      V_i=receiveDisconnectNeighborDegree();
9      if the degree of nodes in V_i are all zero && P_i>0.9
10       equal connection possibility;
11     else if P_i>0.9
12       repeat for all v_j in V_i do
13         calculating connection possibility by degree;
14       until all v_j in V_i are selected
15     use Roulette method select m nodes in V_i base on
       connection possibility;
16     Lst_i = Modify neighbor list of v_i;
17     broadcastEndPacket(Lst_i);
18   until all v_i in {V-V_in} are selected
19 end.
```

This algorithm works as follows. At the beginning, a new joined node i sends an adding connection request to all neighbors. Then, it receives the degree information of nodes which in the local-world of node i. During the adding connection process of node i, the other neighbor nodes cannot execute at the same time. Next, node i calculates connection probability of each neighbors based on feedback information. If the degree of each neighbor node is zero, node i connects with them in equal probability. In other case, node i calculates connection probability according to the degree of neighbor nodes. Then we describe the process of establishing m edges between node i and its neighbors by roulette method. At last, the neighbor list Lst_i is updated and broadcasted to all neighbor nodes.

5   Simulation Results

We evaluate the scale-free properties of our proposed model in Matlab. Total nodes based on homogeneous networks are randomly deployed in a 3D sensor field of $50 * 50 \ m^2$. The number of peaks is set to 12 and their heights are in the range of 10 to 20. Considering each node must have sufficient neighbors during the modeling process, the communication radius r is set to 15 m after many simulations and analysis. The number of nodes in initial wireless sensor network topology N_0 is set to 3 and the maximum degree of all nodes is set to 20.

Figure 3 shows the scale-free properties of network topologies with 100 nodes. All of the nodes are represented by the blue dots. The size of dot is used to distinguish different nodes degree. So bigger dot means higher degree. The black solid lines are used to represent the connection between nodes. In Fig. 3a, the probability parameters of deployed nodes α, β are set to 2 and 0.3. It shows the perceived probability of the deployed sensor nodes. The blue dots represent the projective position of nodes in xy plane. We can find that the majority parts of 3D terrain are achieved over 0.8 perceived probability. They can be monitored with high probability. But there are also some parts with sparse nodes have lower perceived probability. Figure 3b show the scale-free network topology without 3D terrain. Each node needs to add 2 edges during the modeling process and the

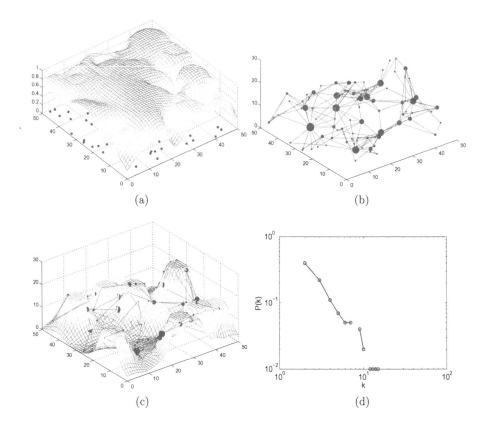

Fig. 3. Scale-free properties of wireless sensor network topologies with 100 nodes in 3D terrain.

perceived probabilities of them are all over 0.9. The majority nodes have low degree. But there are also few nodes with high degree, they have more edges. Nodes degree completely accords with power-law distribution. It proofs that the wireless sensor network topology, which is modeled by the proposed strategy, maintains good scale-free property. In Fig. 3c, we put the completed scale-free network topology in the 3D terrain. The peaks hid some nodes from view. But the location and connection of all nodes keep the same with Fig. 3b. And Fig. 3d shows the degree distribution. It follows the power-law distribution.

6   Conclusion

After fully considering the requirements of wireless sensor networks in practical application, this paper builds a network model with scale-free property based on developed growth and preferential attachment criteria for homogeneous wireless sensor network in 3D terrain. The simulation results show that the scale-free wireless sensor network topology according to our modeling strategy keeps obvious scale-free property.

We will further focus on the application of complex network theory in sensor networks of Internet of Things (IoTs). Combining with the advantages of scale-free model and small world model, we will explore the robust optimization strategy of sensor networks for distributed IoTs in the future work.

Acknowledgments. This work is partially supported by Natural Science Foundation of P.R. China (Grant Nos. 61572106 and 61202443), and the Fundamental Research Funds for the Central Universities.

References

1. Munir, A., GordonRoss, A., Ranka, S.: Multi-core embedded wireless sensor networks: architecture and applications. IEEE Trans. Parallel Distrib. Syst. **25**(6), 1553–1562 (2014)
2. Ji, S., Raheem, B., Cai, Z.: Snapshot and continuous data collection in probabilistic wireless sensor networks. IEEE Trans. Mob. Comput. **13**(3), 626–637 (2014)
3. Dong, H., Zhang, K., Zhu, L.: An algorithm of 3D directional sensor network coverage enhancing based on artificial fish-swarm optimization. In: Proceedings of International Workshop on Microwave and Millimeter Wave Circuits and System Technology, Chengdu, China. pp. 139–142, 19–20 April 2012
4. Hou, J., Wang, J.: The research of enhancing the coverage rate on 3D wireless sensor network. In: Proceedings of International Conference on Computer Science and Education, Melbourne, VIC, Australia, pp. 1281–1284, 14–17 July 2012
5. Lu, S., Wang, Z., Wang, Z., Zhou, S.: Throughput of underwater wireless ad hoc networks with random access: a physical layer perspective. IEEE Trans. Wireless Commun. **14**, 6257–6268 (2015)
6. Pasqualetti, F., Zampieri, S., Bullo, F.: Controllability metrics, limitations and algorithms for complex networks. IEEE Trans. Control Netw. Syst. **1**(1), 40–52 (2014)
7. Hurlburt, G.F.: Complexity theory: a new paradigm for software integration. IT Prof. **15**(3), 26–31 (2013)
8. Watts, D.J., Strogatz, S.H.: Collective dynamics of 'small-world' networks. Nature **393**(6684), 440–442 (1998)
9. Barabási, A.L., Albert, R.: Emergence of scaling in random networks. Science **286**(5439), 509–512 (1999)
10. Colman, E.R., Rodgers, G.J.: Complex scale-free networks with tunable power-law exponent and clustering. Phys. A Stat. Mech. Appl. **392**(21), 5501–5510 (2013)
11. Singh, S., Andrews, J.G.: Joint resource partitioning and offloading in heterogeneous cellular networks. IEEE Trans. Wireless Commun. **13**(2), 888–901 (2014)
12. Jeon, S.W., Devroye, N., Vu, M., Chung, S.Y., Tarokh, V.: Cognitive networks achieve throughput scaling of a homogeneous network. IEEE Trans. Inf. Theory **57**(8), 5103–5115 (2011)
13. Li, R., Yu, J.X., Huang, X., Cheng, H., Shang, Z.: Measuring robustness of complex networks under MVC attack. In: Proceedings of the 21st ACM International Conference on Information and Knowledge Management, New York, NY, USA, pp. 1512–1516, 29 Ocotber–2 November 2012
14. Mahmoud, M.S., Sabih, M.: Experimental investigations for distributed networked control systems. IEEE Syst. J. **8**(3), 717–725 (2014)

15. Liu, L., Qi, X., Xue, J., Xie, M.: A topology construct and control model with small-world and scale-free concepts for heterogeneous sensor networks. Int. J. Distrib. Sens. Netw. **2014**, 1–8 (2014)
16. Du, W., Wu, Z., Cai, K.: Effective usage of shortest paths promotes transportation efficiency on scale-free networks. Phys. A Stat. Mech. Appl. **392**(17), 3505–3512 (2013)
17. Zheng, G., Liu, S., Qi, X.: Scale-free topology evolution for wireless sensor networks with reconstruction mechanism. Comput. Electr. Eng. **38**(3), 643–651 (2012)
18. Zheng, G., Liu, Q.: Scale-free topology evolution for wireless sensor networks. Comput. Electr. Eng. **39**(6), 1779–1788 (2013)
19. Jian, Y., Liu, E., Wang, Y., Zhang, Z., Lin, C.: Scale-free model for wireless sensor networks. In: Proceedings of 2013 IEEE Wireless Communications and Networking Conference (WCNC), Shanghai, China, pp. 2329–2332, 7–10 April 2013

Service Model and Service Selection Strategies for Cross-regional Intelligent Manufacturing

Xinye Chen[✉], Ping Zhang, Weile Liang, and Fang Li

College of Computer Science and Engineering,
South China University of Technology, Guangzhou, China
i_lvorange@163.com

Abstract. Manufacturing plays an important role in the development of national economy. Intelligent manufacturing is the integration of modern business management model, information technology and the traditional manufacturing industry. This paper proposes to handle the following problems that intelligent manufacturing industry faced: (1) the problem of classification and composition of services of different granularity level, (2) the achieve of intelligent logistics over the cross-regional production process, (3) the need of an efficient and stable algorithm for services selection. To achieve cross-regional logistics intelligently and to provide service more intelligently, this paper classify services into five categories according to the level of granularity, and logistics services as the sixth. We design a service ontology based on owl. At last, we design a service selection mode based on the descent of granularity level and a peer service priority-based algorithm for service selection.

Keywords: Service model · Intelligent manufacturing · Web ontology language

1 Introduction

The development trend of manufacturing is intelligent manufacturing. By using artificial intelligence techniques in manufacturing processes, activities such as analysis, reasoning, judgment, decision-making can be carried out by the intelligent system rather than people. So as to solve the problem of labor shortage. With the development of intelligent manufacturing, network is connected with manufacturing. The concept of cloud manufacturing was first proposed in 2009 [1]. Cloud manufacturing is a new networked intelligent manufacturing mode that is service-oriented, low-efficient and knowledge-based. Its manufacturing resources and capabilities are virtualization and service-oriented. The cloud platform centralized these resources and capabilities and provide users with ubiquitous, ready access manufacturing services. Resources sharing and collaborative during the whole life cycle of production. [2, 3] The cloud-based design and manufacturing systems encapsulated the distributed manufacturing resources into manufacturing services. The whole life cycle is composed of these services. Jin et al. and Wei et al. studied on service management for automotive industry in [4, 5]. Wang et al. explore the service modes of optical elements in [6]. Chen discussed the problem of promotion and sustainability of the competitiveness of

© ICST Institute for Computer Sciences, Social Informatics and Telecommunications Engineering 2016
J. Wan et al. (Eds.): Industrial IoT 2016, LNICST 173, pp. 211–221, 2016.
DOI: 10.1007/978-3-319-44350-8_21

enterprises by cloud manufacturing in case of a semiconductor manufacturer in [7]. [8] specialized in the collaboration of design services in cloud manufacturing. Cheng et al. studied on the business model and transaction process in cloud manufacturing in [9]. In order to solve the search and access problem of different manufacturing resources, Tao et al. design a five-layer system in [10], including the resource layer, the perception layer, the network layer, the service layer and the application layer. The system is designed based on things technology. Wang proposed a multi-layer architecture with the use of the standard of IEC 61499 function block in [11]. The system can monitor the status and availability of devices, and use a closed-loop flow of information for process planning. Liu et al. studied on the machining of complex parts in cloud manufacturing and found that some core competences of SPs like know-hows are usually unshareable in [12]. To solve the problem, it encapsulate all services provided by the same SP with standardized machining task description strategies, only the capability information of a service can be provided in the cloud.

Through analyzing the current manufacturing industry, studying on service modes, Cao et al. proposed a working procedure priority-based algorithm (WPPBA) in [13]. The prime collaboration mode is proposed as part manufacturing service combined with working procedure manufacturing service. WPPBA is designed for the selection and composition of working procedure granularity services. Liu proposed a social learning optimization algorithm based on the improved social cognition (SCO) algorithm and improved differential evolution (DE) algorithm and apply to the composition of qos-aware cloud services in [14]. V.Gabrel studied on the composition of qos-aware web services and proposed a mixed integer linear program to represent the problem in [15]. Liu et al. studied on the multi-task oriented services composition to meet high demand of qos of users in [16]. Huang designed a chaos control optimization algorithm for the composition of qos-based services in [17].

Task scheduling is particularly important for intelligent manufacturing. Vahit studied on multi-objective flexible job-shop, and presented an object-oriented (OO) approach in [18]. Many-to-many associations between operations and machines are transformed into two one-to-many associations.

The rest of this paper is organized as follows: Sect. 2 discuss the granularity of services in the system. Section 3 presents the semantic descriptions of the services and the tasks. Section 4 presents the problem definition for service selection. Section 5 presents the experimental results; Sect. 6 summarizes the paper.

2   Granularity of Services

Many articles classified manufacturing services into four categories, products, components, parts and working procedure, according to their granularity. A product is composed of some components, a component is composed of some parts, a part is composed of some working procedures. That is, a product is not only composed of parts, but also working procedures. Cao et al. [13] focus on the granularity of parts and working procedure. Outsourcing would be used on the condition of having not enough time, equipment failure or having some demand beyond the capacity of the Enterprise. The outsourcing part is a part of the production of the enterprise. So the granularity of the outsourcing part should be smaller than that of the whole production. Users may have no idea what services of lower level of granularity to apply for in case of they don't know the specific processes of the service of a higher level of granularity. So, all of the four level of granularity of services should be involved in the system. There are two types of service in each level, design services and manufacturing services. A design service describe the specific processes of the service composed of services in lower level of granularity.

Hardware service and software service should also involved in the system to ensure that the system can meet a variety of manufacturing requirements. The level of granularity of hardware and software services are considered as the level of tools. One other indispensable kind of service in cross-regional intelligent manufacturing is transport service. Transport service is a little different from other services, because it's carried on during the interval of two services and its selection has nothing to do with the user's demand but the location of the two selected services before and after it. Each level is possible for the two services. Therefore, the authors believe that services in cross-regional intelligent manufacturing should be divided into six categories:transportation services, production services, component services, parts services, working procedure services and tools services.

3 The Model of Cross-regional Intelligent Manufacturing Service

CIMS presents the cross-regional intelligent manufacturing service class. It has seven sub-classes: WPIMS (working procedure service), PIMS (parts service), CPIMS (components service), PRIMS (product service), SWIMS (software service), HWIMS (hardware service), TRIMS (transportation service). CIMT presents the cross-regional intelligent manufacturing task class;

$$CIMS = \{GeneralInfo, AbilityInfo, QosInfo, Process, State, Resource\} \quad (1)$$

$$CIMT = \{ID, Process, start, complete\} \quad (2)$$

1. GeneralInfo presents the basic infomation of services:

$$GeneralInfo = \{ID, Name, Description, Scale, Type, Category, ContractInformation\} \quad (3)$$

ID is used to identify the services;Name is related to the specific content of the service;Description describe the content of service in more detail;Scale presents whether the service can sell in small batch, large quantities or retail;Type presents the granularity of the service;Category presents the function type of the service, so that the service selecting can be more efficient; ContractInformation describe the contact information of the service provider, such as address, phone, email-address, etc.
2. AbilityInfo describe the information on the capacity of a service. Each sub-class of CIMS related to a different sub-class of AbilityInfo. Each of the sub-class of AbilityInfo describe the quality, the optional size, the variable parameters and the invariable parameters. AbilityInfo of transportation service should describe its

service area. AbilityInfo of hardware service describe its speed and machining range. AbilityInfo of software service describe its calculation range and so on.

3. QosInfo presents the Qos values of the service. These value is needed to calculate the selecting value while selecting services.

$$QosInfo = \{price, efficiency, evaluation\} \tag{4}$$

price presents the unit price of a service;efficiency presents the number of the service can be provided per unit of time;evaluation present the average of the history evaluation of the service.

4.
$$State = \{Availability, task\} \tag{5}$$

Availability presents the status of the service. The status may be iddle, occupied or out of service;task indicates the list that the service has been scheduled. The calculation of the value of waiting time while selecting services is based on this.

5. Resource describes the resource information of a service:

$$Resource = \{Material, MaterialResource, HumanResource, Equipment, Enterprise\} \tag{6}$$

Material presents materials the machining required;MaterialResource represent the specific amount of material required for unit service;Equipment describe the devices used in the service;Enterprise represents the company providing the service.

6. Process describe the processing of the service, including the construction information of sub-services:

7.
$$Process = \{Input, output, startTime, ControlConstruct\} \tag{7}$$

Input describes the needed inputs of the service; output describes the outputs; startTime describes the relative start time of ControlConstruct to the start of the service; ControlConstruct represents the logical relationships between services, with reference to the owl-s. It is facilitate to the decomposition of the task.

$$ControlConstruct = (Sequence|split|split + join|Any - Order|Choice|If - \atop Then - Else|Iterate|Repeat\_While|Repeat\_Until|Perform|Produce) \tag{8}$$

Perform presents a sub-service in the construction. This service may correspond to a specific service in the system, an abstract sub-service represent the process of the service, or a undetermined service waiting to select a specific one.

$$Perform = \{Service, Category, type, transport\} \tag{9}$$

If the Perform does not present a specific service, it should present a Category;type describe the service is an specific one or an abstract one;If there is a need for transportation after the service, transport describe the transportation service.

4 Peer Services Priority-based Algorithm

4.1 Problem Definition

In accordance with the service selecting mode described in the previous chapter, the selection of services of each level of granularity is similar. It is necessary to take into account both the time and the cost while selecting services. The notations used in definition are as in [13] as follows:

P_i	The ith part-level manufacturing task, $i = 1, 2, ..., k$
$wp_{i,j}$	The jth WP of P_i, $j = 1, 2, ..., n(i)$
$s_{i,j}$	The machining method of $wp_{i,j}$
$t_{i,j}$	The working hours of $wp_{i,\ j}$
RP_i	The ith RP in the system, $i = 1, 2, ..., h$
M_i	The ith machining method considered in this model, $i = 1, 2, ..., e$
$c(i, j)$	The price per hour of M_j that provided by RP_i
$\bar{c}(j)$	The average price per hour of M_j available in all RP_s
$\delta(j)$	The variance of price per hour of M_j available in all RP_s
$d(i, j)$	Logistics time between RP_i and RP_j
$lc(i, j, m)$	Logistics cost between RP_i and RP_j when the delivered weight is m
m_i	The workblank mass of P_i
A_i	The workblank supplier of P_i is in the same city as A_ith RP
B_i	The SD of P_i is in the same city as B_ith RP
$w(i, j)$	The waiting time of service M_j in RP_i
X_i	The processing route of P_i, $Xi = [x_i,1, x_i,2, ..., x_i,n(i)]$
$TC_i/MC_i/LC_i$	The total cost/machining cost/logistics cost of P_i
$TT_i/MT_i/LT_i/WT_i$	The total time/machining time/logistics time/waiting time of P_i
α_i/β_i	The cost weighting factor/time weighting factor of P_i
$mc(i)$	The largest acceptable machining price per hour of M_i
AC_i/AT_i	The largest acceptable total cost/longest acceptable total time of P_i

TC_i, MC_i, LC_i, TT_i, MT_i, LT_i, WT_i are calculated as follows [13]:

$$\begin{cases} TC_i = MC_i + LC_i \\ MC_i = \sum_{j=1}^{n(i)} c(x_{i,j}, s_{i,j}) t_{i,j} \\ LC_i = \delta m_i \left(d(A_i, x_{i,1}) + \sum_{j=1}^{n(i)-1} d(x_{i,j}, x_{i,j+1}) + d(x_{i,n(i)}, B_i) \right) \end{cases} \quad (10)$$

$$\begin{cases} TT_i = MT_i + LT_i + WT_i \\ MT_i = \sum_{j=1}^{n(i)} t_{i,j} \\ LT_i = 24\left(d(A_i, x_{i,1}) + \sum_{j=1}^{n(i)-1} d(x_{i,j}, x_{i,j+1}) + d(x_{i,n(i)}, B_i)\right) \\ WT_i = 24 \sum_{j=1}^{n(i)} w(x_{i,j}, s_{i,j}) \end{cases} \tag{11}$$

The mathematical model could be given as follows [13]:

$$\min \sum_{i=1}^{k} f_i(X_i) = \min \sum_{i=1}^{k} (\alpha_i \bullet TC_i + \beta_i \bullet TT_i) \tag{12}$$

$$s.t. \begin{cases} c(x_{i,j}, s_{i,j}) < mc(s_{i,j}) \\ TC_i < AC_i \\ TT_i < AT_i \\ \alpha_i + \beta_i = 1 \end{cases}$$

The problem is to minimizing the total cost and total time. The constraints of AC_i and AT_i can be met by adjusting the weighting factors α_i and β_i [13] if it is possible.

4.2   Algorithm Design

The most widely used service selection algorithms are genetic algorithm, differential evolution algorithm, particle swarm optimization and colony optimization, etc. But the drawback of these intelligent algorithms is can not ensure rapid responses [13]. Yang et al. [13] proposed a WP priority-based algorithm (WPPBA). Major WP_s is determined by $c(s_{i,j}, j)t_{i,j}$. The sum of $c(s_{i,j}, j)t_{i,j}$ of all the major WP_s should be more than 75 % of the total machining cost. The minor WP_s between the major WP_s should select the same SP as the left major one or the right major one. So that the logistics cost would be minimized. But the algorithm is not stable enough under some special circumstance.

Consequently, inspired by the WPPBA, we propose a peer services priority-based algorithm (PSPBA) which is more stable. We use $dx(j)$ as the key factor to determine the major services. The major define value of each service is as follow:

$$MDV_{i,j} = dx(j)t_{i,j}^2 \tag{13}$$

Sort the services in descending order of $MDV_{i,j}$:

$$so = \left\{wp_{i,a}, wp_{i,b}, wp_{i,c}, \cdots \mid MDV_{i,a} > MDV_{i,b} > MDV_{i,c} > \ldots\right\} \tag{14}$$

If $wp_{i,j}$ is in the top 30 % of SO, or

$$MDV_{i,j} > 4\bar{d}_i^2 \tag{15}$$

$$\bar{d}_i = average\{d(j,k)|j,k<n_i\} \tag{16}$$

than, $wp_{i,j}$ is one of the main services.

The sequence of the selection of major services follow the services order:

$$\min(\alpha_i(c(x_{i,j},s_{i,j})t_{i,j} + \delta m_i(d(last,x_{i,j}) + d(x_{i,j},next))) + \\ \beta_i(t_{i,j} + 24(d(last,x_{i,j}) + d(x_{i,j},next) + w(x_{i,j},s_{i,j})))) \tag{17}$$

$last$ indicates $x_{i,k}$ of the nearest service before this one and $x_{i,k} \neq 0$, which means $wp_{i,k}$ have been selected. *Otherwise, last* $= A_i$next indicates $x_{i,k}$ of the nearest service after this one and $x_{i,k} \neq 0$. *Otherwise, last* $= B_i$.

The selection of the minor services between major services refers to WPPBA [13], the processing route of between $wp_{i,p}$ and $wp_{i,q}$ is one column of the matrix $X_{p,q}$:

$$X_{p,q} = \begin{pmatrix} x_{i,p} & x_{i,p} & x_{i,p} & x_{i,p} & \cdots & x_{i,p} & x_{i,p} & x_{i,q} \\ x_{i,p} & x_{i,p} & x_{i,p} & x_{i,p} & \cdots & x_{i,p} & x_{i,q} & x_{i,q} \\ x_{i,p} & x_{i,p} & x_{i,p} & x_{i,p} & \cdots & x_{i,q} & x_{i,q} & x_{i,q} \\ \vdots & \vdots & \vdots & \vdots & \ddots & \vdots & \vdots & \vdots \\ x_{i,p} & x_{i,p} & x_{i,p} & x_{i,q} & \cdots & x_{i,q} & x_{i,q} & x_{i,q} \\ x_{i,p} & x_{i,p} & x_{i,q} & x_{i,q} & \cdots & x_{i,q} & x_{i,q} & x_{i,q} \\ x_{i,p} & x_{i,q} & x_{i,q} & x_{i,q} & \cdots & x_{i,q} & x_{i,q} & x_{i,q} \end{pmatrix} \tag{18}$$

To determine the process route between $wp_{i,p}$ and $wp_{i,q}$ is to find the column:

$$\min(\alpha_i \sum_{j=p}^{q} c(x_{i,j},s_{i,j})t_{i,j} + 24\beta_i \sum_{j=p}^{q} w(x_{i,j},s_{i,j})) \tag{19}$$

The calculate process of peer service priority-based algorithm is described as follows:

Step 1 Calculate $MDV_{i,j}$ according to Formula (4).

Step 2 Sort the services and get major services based on Formula (5)-(7).

Step 3 Select each major service follow the order of (5) and Formula (8).

Step 4 Optimize the processing route of minor WP_s based on Formula (9)-(10).

Step 5 Check whether the selected services meet the requirements. Otherwise, adjust the value of α_i and β_i, or adjust the imperfect service to another RP.

5 Simulation Experience and Evaluation

5.1 Initial Data

To prove the practicability of PSPBA in selecting services, we take ten customized mechanical parts as test cases. Eight of the ten parts have been used to test the

Table 1. Basic information of tasks.

part	m (kg)	A	B	$x_{i,j}/t_{i,j}$ $w_{i,1}$	$w_{i,2}$	$w_{i,3}$	$w_{i,4}$	$w_{i,5}$	$w_{i,6}$	$w_{i,7}$	$w_{i,8}$	$w_{i,9}$	$w_{i,10}$
P_1	8.14	2	4	1/3.55	3/2.55	7/0.50	4/0.17	5/2.46	2/0.50	9/0.21	7/0.60	–	–
P_2	9.86	5	1	1/1.96	3/0.80	4/0.07	1/0.75	5/0.54	4/0.04	5/0.48	6/0.05	2/0.82	7/0.33
P_3	4.98	6	7	2/1.30	9/0.87	7/0.42	4/0.10	5/2.78	4/0.17	7/0.50	–	–	–
P_4	5.28	3	6	1/1.37	4/0.57	6/0.08	7/0.50	1/0.61	5/2.74	2/0.07	3/0.17	7/0.45	–
P_5	2.88	2	5	8/0.27	4/0.18	6/0.05	8/0.43	4/0.10	5/1.95	7/0.25	–	–	–
P_6	4.46	4	3	1/0.61	3/0.75	4/0.17	6/0.10	5/2.35	9/0.90	7/0.42	–	–	–
P_7	2.62	1	5	1/0.48	5/0.53	1/0.32	3/0.70	5/1.06	4/0.43	3/0.06	7/0.17	–	–
P_8	4.40	4	2	2/0.21	5/3.08	4/0.23	6/0.37	2/0.35	7/0.17	3/0.75	7/0.15	–	–
P_9	5.99	3	3	1/0.61	3/0.75	7/0.64	4/0.10	5/0.48	4/0.04	5/0.54	6/0.43	2/0.82	7/0.45
P_{10}	3.06	2	6	1/3.55	3/2.44	4/0.04	1/0.63	5/0.51	4/0.67	5/0.47	2/0.05	9/0.48	7/0.27

performance of WPPBA by Yang et al. [13]. We cited these eight sets of data to test the performance of PSPBA, and make a comparison between WPPBA (Table 1).

Nine kinds of machining methods and seven RP_s are include in the simulation experiments. c is the price per hour matrix (yuan/h), whose rows denote RP and columns denote machining methods; d is the RP distribution matrix (day); w is the waiting time matrix(day), whose rows denotes machining methods and columns denotes RP_s. They are formulated as below [13]:

$$c = \begin{pmatrix} 36 & 36 & 41 & 16 & 48 & 36 & 999 & 103 & 109 \\ 999 & 999 & 28 & 32 & 38 & 34 & 36 & 88 & 90 \\ 23 & 44 & 33 & 34 & 49 & 26 & 24 & 100 & 98 \\ 38 & 28 & 43 & 29 & 36 & 39 & 30 & 100 & 87 \\ 33 & 44 & 41 & 30 & 41 & 21 & 999 & 85 & 103 \\ 27 & 999 & 44 & 30 & 36 & 29 & 33 & 84 & 88 \\ 26 & 35 & 38 & 999 & 37 & 28 & 34 & 95 & 95 \end{pmatrix}$$

$$d = \begin{pmatrix} 0 & 2 & 2 & 2 & 3 & 2 & 3 \\ 2 & 0 & 1 & 1 & 2 & 2 & 3 \\ 2 & 1 & 0 & 1 & 2 & 2 & 3 \\ 2 & 1 & 1 & 0 & 2 & 2 & 3 \\ 3 & 2 & 2 & 2 & 0 & 2 & 3 \\ 2 & 2 & 2 & 2 & 2 & 0 & 2 \\ 3 & 3 & 3 & 3 & 3 & 2 & 0 \end{pmatrix}$$

$$w = \begin{pmatrix} 3 & 2 & 3 & 2 & 3 & 2 & 1 & 4 & 2 \\ 2 & 2 & 2 & 2 & 2 & 3 & 2 & 4 & 4 \\ 4 & 4 & 3 & 3 & 5 & 1 & 2 & 4 & 2 \\ 3 & 3 & 3 & 1 & 2 & 2 & 4 & 3 & 2 \\ 5 & 3 & 1 & 5 & 1 & 5 & 1 & 3 & 4 \\ 2 & 2 & 2 & 5 & 4 & 5 & 1 & 2 & 4 \\ 2 & 4 & 1 & 3 & 3 & 3 & 2 & 4 & 1 \end{pmatrix}$$

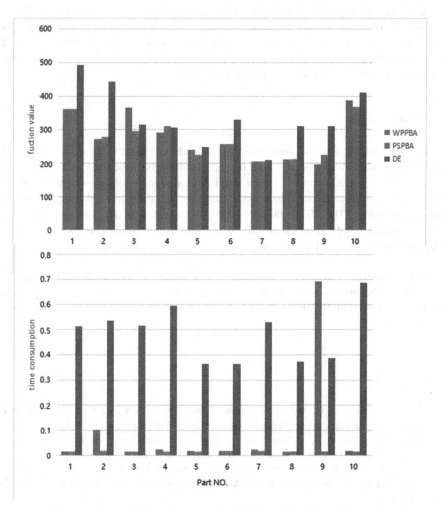

Fig. 1. Time consumption and function value comparison of PSPBA, WPPBA and DE

5.2   Experiment Results

Standard DE and WPPBA [13] are selected for comparison to verify the effectiveness of PSPBA. The population size of DE is 20; crossover probability is 0.85; and the iteration number is 100. Figure 1 shows the function value and the time consumption of the ten sets of solution of the three algorithms.

Obviously, if a lower function value denotes a better solution, the solution quality of DE is much lower than that of WPPBA and PSPBA, and the solution quality of WPPBA is a litter lower than that of PSPBA. However, in practice, time consumption is also important. Time consumption of PSPBA is much more stable than WPPBA. Time consumption of WPPBA is affected by the numbers of its major services. When the number of major services increase, time consumption would rise exponentially. As the result shows, selecting major services one by one with influence of the selection of nearby services does not greatly affect the solution quality. What's more, time consumption would be reduced greatly, and more stable.

6   Conclusion

A way of classification and composition of manufacturing services in cross-regional intelligent manufacturing is proposed. Information models of the services and tasks are constructed. Then a algorithm called PSPBA is designed for services selection in intelligent manufacturing to minimize the total cost and time of a task, and reduce the time consumption. The effectiveness of PSPBA is validated by simulation experiments. The simulation results shows that PSPBA has less time consumption and better solution compared with WPPBA and DE.

Acknowledgments. This paper is supported by Science and Technology Planning Project of Guangdong Province, China (2014B090921007), Science and Technology Program of Guangzhou, China (20150810068), Science and Technology program of Haizhu District, China (2014-cg-02).

References

1. Bo-hu, L.I., Zhang, L., Wang, S.L., et al.: Cloud manufacturing: a new service-oriented networked manufacturing model. J. Comput. Integr. Manufact. Syst. **16**, 1–1343 (2010)
2. Liu, Q., Wan, J., Zhou, K.: Cloud manufacturing service system for industrial-cluster-oriented application. J. Internet Technol. **15**(3), 373–380 (2014)
3. Wang, S., Wan, J., Li, D., Zhang, C.: Implementing smart factory of industrie 4.0: an outlook. Int. J. Distrib. Sens. Netw. **2016**, 10 (2016). doi:10.1155/2016/3159805. Article ID 3159805
4. Wei, C.M., Zhang, C.L., Song, T.X., Huang, B.Q.: A cloud manufacturing service management model and its implementation. In: Proceedings of International Conference on Service Sciences, vol. 3, pp. 60–63 (2013)

5. Jin, Z.X.: Research on solutions of cloud manufacturing in automotive industry. In: SAE-China, FISITA (eds.) Lecture Notes in Electrical Engineering, vol. 199, pp. 225–234 (2013)
6. Wang, J., Wang, J., Tao, Y.Z., Li, H.M., Li, W., Jiang, L.Q.: The exploration of cloud manufacturing service mode for high power laser optical elements. In: Proceedings of the SPIE International Society Optical Engineering (2012)
7. Chen, T.: Strengthening the competitiveness and sustainability of a semiconductor manufacturer with cloud manufacturing. J. Sustain. **6**, 251–266 (2014)
8. Liu I, Jiang H.: Research on key technologies for design services collaboration in cloud manufacturing. In: 16th IEEE International Conference on Computer Supported Cooperative Work in Design (CSCWD), pp. 824–829 IEEE (2012)
9. Cheng, Y., Zhang, Y., Lv, L., Liu, J.R., Tao, F., Zhang, L.: Analysis of cloud service transaction in cloud manufacturing. In: IEEE International Conference on Industrial Informatics, pp. 320–325 (2012)
10. Tao, F., Zuo, Y., Xu, L.D., et al.: IoT-based intelligent perception and access of manufacturing resource toward cloud manufacturing. J. IEEE Trans. Ind. Inform. **10**, 1547–1557 (2014)
11. Wang, L.: Machine availability monitoring and machining process planning towards Cloud manufacturing. CIRP J. Manufact. Sci. Technol. **6**, 263–273 (2013)
12. Liu, X., Li, Y., Wang, L.: A cloud manufacturing architecture for complex parts machining. J. Manufact. Sci. Eng. **137**, 061009 (2014)
13. Cao, Y., Wang, S., Kang, L., et al.: Study on machining service modes and resource selection strategies in cloud manufacturing. Int. J. Adv. Manufact. Technol. pp. 1–17 (2015)
14. Liu, Z.Z., Chu, D.H., Song, C., et al.: Social learning optimization (SLO) algorithm paradigm and its application in QoS-aware cloud service composition. Inf. Sci. Int. J. **326**, 315–333 (2016)
15. Gabrel, V., Manouvrier, M., Murat, C.: Web services composition: complexity and models. J. Discrete Appl. Math. **196**, 67–82 (2014)
16. Liu, W., Liu, B., Sun, D., et al.: Study on multi-task oriented services composition and optimisation with the'multi-composition for each task' pattern in cloud manufacturing systems. Int. J. Comput. Integr. Manufact. **26**, 786–805 (2013)
17. Huang, B., Tao, C.L.F.: A chaos control optimal algorithm for QoS-based service composition selection in cloud manufacturing system. J. Enterp. Inf. Syst. **8**, 445–463 (2014)
18. Vahit, K.: Lu V. An object-oriented approach for multi-objective flexible job-shop scheduling problem. Expert Syst. Appl. Int. J. **45**, 71–84 (2016)

A Model-Based Service-Oriented Integration Strategy for Industrial CPS

Fang Li[✉], Ping Zhang, Hao Huang, and Guohao Chen

School of Computer Science and Engineering, South China University of Technology,
Guangzhou, China
cslifang@scut.edu.cn

Abstract. In order to realize efficient and unambiguous development of reliable industrial CPS, we employ a model-based service-oriented integration approach, which adopts a model-centric way to automate the development course of the entire software life cycle. The structures and rules for iCPS modeling and hierarchical modeling elements are defined in the meta-model, including services, and function blocks of different abstraction level. The relationship between service and function blocks are also defined clearly in the meta-model. A UML-compliant graphical modeling environment is generated from the meta-model, with a suite of fully integrated tools. The approach is then used to develop the industrial assembly line system. It is an attempt to support iCPS design in an effective way, at the same time guarantee the system performance requirements.

Keywords: Industrial cyber-physical system · Model-based · Service-oriented · Integration

1 Introduction

The advances in information and computer technology have led to the improved integration of heterogeneous devices and systems in industrial automation. This has also led to a new type of industrialization named industry 4.0 in Germany [1, 2]. Industrialists, researchers and practitioners are focusing their eyes on the develop of a set of large process industry systems, in the form of industrial Cyber-Physical System(iCPS), in which a collection of devices are interconnected and communicating with each other by networks.

Designing and developing iCPS means to deal with several challenges for handling complexity of the system. From the functional point of view, one of the major challenges is focused, on one side, on managing the constantly increasing integrated functions and the vastly increased number of devices integrated from different manufacturers. Take the industrial assembly line as an example, not only various technologies are integrated in such system, such as, motion control, visual inspection, wireless sensing etc, but also several devices such as robotics, PLC, IPC, RFID etc. are integrated make it a large-scale control and monitoring system. On the other side, such heterogeneous systems combine mechanical components, embedded systems, networking, application software

© ICST Institute for Computer Sciences, Social Informatics and Telecommunications Engineering 2016
J. Wan et al. (Eds.): Industrial IoT 2016, LNICST 173, pp. 222–230, 2016.
DOI: 10.1007/978-3-319-44350-8_22

and user interfaces to interact with each other. Crosscutting concerns from the multi-disciplinary domain make the integration of the system be a complexity task. From the non-functional point of view, rigorous performance demands are also required, for example, real-time performance and safety criticality, improved reliability and predictability, modularity, flexibility, reusability and reconfigurability, associated with design, test and verification of end products.

MBD(Model-based design) [3–6] and CBD(Component-based design) [7–10] have been proposed as efficient methods in embedded system development, which have also been widely applied in CPS development [11–13]. These approaches are effectively useful in performance guarantee and complex settlement, at the same time in allowing reuse of development efforts by components usage.

Meanwhile, SOA(Service-oriented architecture) [14–16] is being promoted based on CBD, in which, a collection of services along with an infrastructure to enable these services communicate with each other. Services are loosely-coupled and interact based on through internet, which makes SOA more flexible than CBD.

While, as the level of complexity of today's iCPS is continually increasing, problems still need to solve in large-scale complicated system design. We indicate here some key questions will need to be conducted on future industrial CPS systems, including:

1. The lack of group management for devices. As a set of individual devices are interconnected to form networks in iCPS, new ways of easily managing millions of devices in large-scale and complex systems need to be considered.
2. The lack of temporal semantics in service model. As networking technologies make predictable and reliable real-time performance difficult, synchronous and asynchronous communication in the system will be considered.
3. The function-service integration. As service is defined as coarsely-grained functional entity, an interesting question that need to be considered is how the functions will be mapped to services, for smooth integration.

   We have previously proposed a model-based integration approach for CPS, including a CPS integration modeling language and a set of integrated development tools, based on IEC 61499 components [17]. As an extension of the previous work, this paper adopts model-based service-oriented integration approach in industrial CPS development. At first, the model-based service-oriented integration approach is proposed. Then, the hierarchical model definition and meta-model for service-oriented integration modeling are described. At last, a case study for an industrial assembly line system development is conducted to demonstrate the application of the proposed approach. Finally, the concluding section summarizes the features of our work and their implications.

2 The Model-Based Service-Oriented Integration Approach

In order to realize efficient and unambiguous development of reliable industrial CPS, we employ a model-based service-oriented integration approach, which was evolved as an extension of CPS integration framework [18]. The approach adopts a model-centric way to automate the development course of the entire software life cycle, in an iterative and interactive way. Also, the approach integrates a set of tools to support the complete development cycle, including requirement analysis, model specification, simulation & verification, code synthesis as well as implementation.

As illustrated in Fig. 1, a UML-based meta-model is used to define the structures and rules for iCPS modeling. Hierarchical modeling elements are defined in the meta-model, including services, and function blocks of different abstraction level. The interaction and connection between components, the relationship between service and function blocks are also defined clearly in the meta-model. Then, A UML-compliant graphical modeling environment is generated from the meta-model, with a suite of fully integrated tools, i.e. tool adaptors, code generators, et al.

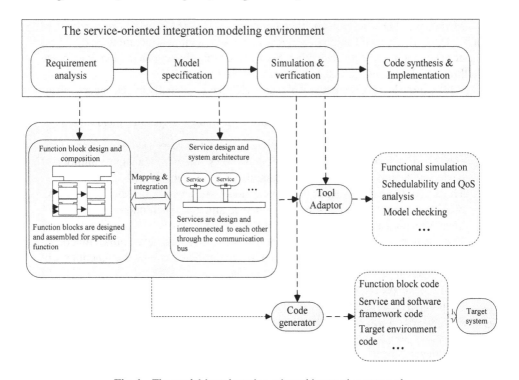

Fig. 1. The model-based service-oriented integration approach

In the application development process, models can be built in the service-oriented integration modeling environment after requirement analysis. It is a hierarchical component-based model with multiple-level of abstraction. In the system layer, services are designed, as a coarsely-grained functional entity that interacts with other services through a loosely-coupled communication model, for example, through a software communication bus. Then, a set of hierarchical function blocks are defined to implement the services, including BFB, CFB, SIFB and FU, partly according to IEC61499 standard. Integration can then be conducted through a mapping from function blocks to services.

Using tool adaptors in the environment, various simulation and verification toolsets (Matlab, UPPAAL et al.) can be integrated, for function and performance simulation and verification to guarantee the function and performance requirements of the system. These tool adaptors extract specific model information and then transform it to different simulation and verification tools. For example, functional simulation, Schedulability and QoS analysis and model checking can be conducted using different tools. Also, It is an iterative process because the verification results can then feedback to tell the developer gives some modifications to model or not.

Engaging code generators, automatic code synthesis can be carried out, by inputting service-oriented integration model, and producing the platform specific configuration and the corresponding codes to build the final executable application. The object executive codes include function block code, services and software framework code and target platform code. Function block code deal with the function execution in the software, which come from pre-constructed function block library. The services and software framework code deal with network communication and connection in the system to ensure the integrity of events interaction as well as the accuracy of data interaction. Target platform code deal with the task implementation details and target environment information.

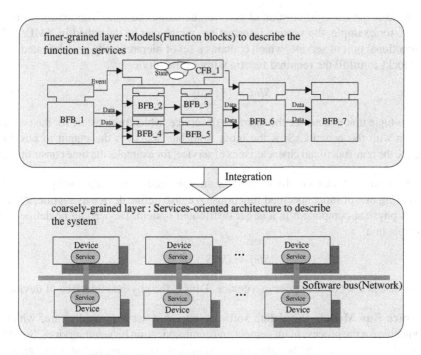

Fig. 2. The service-oriented integration model

3   The Service-Oriented Integration Model for iCPS

We propose a multi-layer integration model to specify iCPS. In the coarsely-grained layer, service oriented architecture is utilized through the application of a universal communication framework, while, in the finer-grained layer defines various kinds of function blocks to describe the system function, as shown in Fig. 2.

3.1 Structural Model of the System

Structurally, a system is conceived as a network of communicating services, which interact transparently with each other by exchanging labeled messages through the communication bus. Collaborating devices are defined in system model as the providers of services.

Service Model. A service is defined as a well-defined, self-contained functional entity that does not depend on the context or state of other services. It is a coarsely-grained functional entity that interacts with other services through a loosely-coupled communication model. Service can be defined as a tuple

$$Service = (S_d, FU, SInf, QoS) \tag{1}$$

where, Sd is the basic description of service, including the basic information as to the service, for example, the version information, which can be described in XML. FU is the functional unit of service, which contains a set of hierarchically prefabricated function blocks to fulfill the required functionality of the service.

$$SInf = (S_i, S_o) \tag{2}$$

is a two-tuple unit to describe the interface of service, which responsible for the communication with the network. Si is the input from bus and so is the output to bus. QoS describe the non-functional characteristics of service, for example, the time constrains, etc.

Device Mode. Device can be defined as an independent physical entity capable of performing one or more specified functions. We also define device as supplier of services. A physical component in a larger distributed system. Device can be defined as a two tuple unit,

$$SInf = (S_i, S_o) \tag{3}$$

where, Rsc is a set of resources in a device. DInf is the physical interface of device.

Software Bus Model. We define software bus as a service middleware, which is responsible for the process management and communication between services. Software bus model include several modules, i.e. communication module, resource management module, data storage module, Scheduling module, etc. The service orchestration can also be organized in software bus.

Function Unit. Function Unit is coarse-grain reusable component to describe the function in a service. It consists of a number of hierarchical connected function blocks, including BFB, CFB, and SIFB, partly according to IEC61499 standard, which is detailed described in [18]. BFB(Basic function block) is regarded as the fundamental function element to construct the system, which cannot be further refined. CFB(Composite function block) is a composition of BFB or other CFB to accomplish a specific complex function. SIFB (Service interface function block) is a type of service interface responsible for communication between inner function block and outer environment.

3.2 The Meta-model for Service-Oriented Integration Modeling

The proposed meta-model defines the system structure, components, as well as relationships and constrains, which include the concepts of embedded control, real-time as well as service-oriented domains. The UML compliant graphic modeling environment allows the modeling of devices, services, function blocks and also their relationships and characteristics. Figure 3 depicts part of the proposed meta-model for service-integration modeling.

As the core modeling elements of the system, a Service has a uniquely defined ServiceName and ServiceID, which can be distinguished from other services. SerDescription and QoS are contained in a Service, represent the basic service description and quality of service. In the QoS, time constrains of a service are defined, for example, the period, priority, WCET(the worst-case execution time) etc. FU is defined as the function unit that contained in a Service to describe the function of service. Services can be connected to software bus through the input and output ports. The Orchestrator would help in orchestrating all services involved in a system.

As a service provider, the device model include a set of resources, in which, some platform attributes are defined, for example, the location, process frequency, process_type, etc.

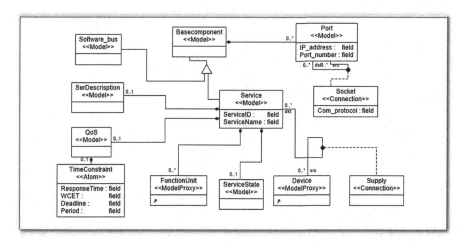

Fig. 3. Part of the meta-model for service-oriented integration modeling

4   A Case Study

4.1   The Assembly Line Description

To demonstrate our approach, we use the industrial assembly line development as a documented example. The industrial assembly line used in this demonstration is composed of feeding robots for automatic upper and lower material and split robot, equipped with IPC for information management and a conveyor controlled by PLC. Sample pieces in the feed box are all labeled with RFID.

In the work flow, the sample pieces are picked up to the conveyor, and information in the RFID label then can be read by sensors. The results are sent to the robot controller to command the robot to pick them from the conveyor and place them into a classified box. Devices in the system include robots, IPC, conveyor, RFID, sensors, et al. These devices are all connected by internet using TCP/IP.

4.2   Service-Oriented Integration Model Definition

The industrial assembly line can be modeled in the UML compliant service-oriented integration environment, as depicted in Fig. 4. A set of services are defined in the model, i.e. Feeding_service, Conveyor_service, PLC_service, etc. These services can be connected through the Software_bus. Service interface and communication protocol can also be defined in the model. Then, these service can be refined by defining a set of communication function blocks to describe the functions.

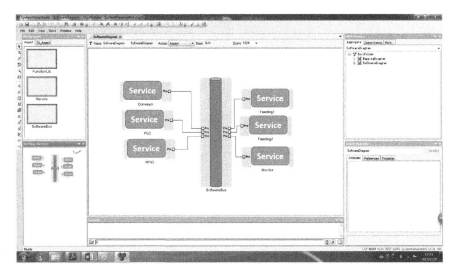

Fig. 4. The application model for industrial assembly line

4.3   The Implementation of the System

After the application model build, the verification and implementation can be conducted based on the integrated toolsets. The generated codes include function block code, services and software framework code and target platform code. Figure 5 depict the generated GUI for service management in the system.

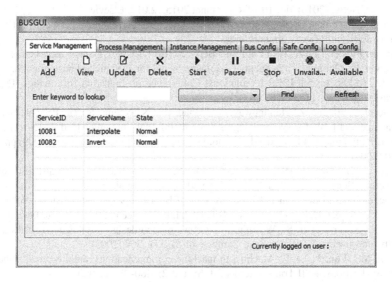

Fig. 5. The generated GUI for service management

5   Conclusion

A service-oriented integration approach for iCPS design and development is presented in the paper. It is a model-centric way to automate the development course of the entire software life cycle. By integrating a set of tools, the complete development cycle, including requirement analysis, model specification, simulation & verification, code synthesis as well as implementation can be conducted in the environment.

The structures and rules for iCPS modeling and hierarchical modeling elements are defined in the meta-model, including services, and function blocks of different abstraction level. The relationship between service and function blocks are also defined clearly in the meta-model. A UML-compliant graphical modeling environment is generated from the meta-model, with a suite of fully integrated tools. The approach is then used to develop the industrial assembly line system. It is an attempt to support iCPS design in an effective way, at the same time guarantee the system performance requirements.

Acknowledgments. This paper is supported by Science and Technology Planning Project of Guangdong Province, China (2014B090921007), Science and Technology Program of Guangzhou, China (20150810068), Science and Technology program of Haizhu District, China (2014-cg-02).

References

1. Wang, S., Wan, J., Li, D., Zhang, C.: Implementing smart factory of industrie 4.0: an outlook. Int. J. Distrib. Sensor Netw. **2016**, 10 p., (2016). doi:10.1155/2016/3159805
2. Wang, S., Wan, J., Zhang, D., Li, D., Zhang, C.: Towards the smart factory for industrie 4.0: a self-organized multi-agent system assisted with big data based feedback and coordination. Comput. Netw. (2016). doi:10.1016/j.comnet.2015.12.017. Elsevier
3. Object Management Group. Model Driven Architecture - A Technical Perspective [EB/OL], 9 July 2001. http://www.omg.org/mda/
4. Ptolemy Project. University of California Berkeley [EB/OL]. http://ptolemy.eecs.berkeley.edu
5. Vyatkin, V., Dubinin, V.: Sequential axiomatic model for execution of basic function blocks in IEC 61499. In: 5th IEEE International Conference on Industrial Informatics (INDIN), Vienna, 2007, pp. 1183–1188 (2007)
6. Porter, J., Lattmann, Z., Hemingway, G., Mahadevan, N., Neema, S.: The ESMoL modeling language and tools for synthesizing and simulating real-time embedded systems. In: 15th IEEE Real-Time and Embedded Technology and Applications Symposium 2009, pp. 112–130 (2009)
7. International Electro-technical Commission (IEC). International Standard IEC 61499, Function Blocks, Part 1–Part 4[S]. IEC Jan 2005. http://www.iec.ch/
8. Function Blocks Development Kit. Holobloc Inc [EB/OL], July 2005. http://www.holobloc.com
9. CORFU Project. University of Patras [EB/OL]. http://seg.ee.upatras.gr/corfu/dev/index.htm
10. Lee, E., Neuendorffer, S., Wirthlin, M.: Actor-oriented design of embedded hardware and software systems. J. Circuits Syst. Comput. **12**(3), 231–260 (2003)
11. Chen, M., Wan, J., Li, F.: Machine-to-machine communications: architectures, standards, and applications. KSII Trans. Internet Inf. Syst. **6**(2), 480–497 (2012)
12. Zou, C., Wan, J., Chen, M., Li, D.: Simulation modeling of cyber-physical systems exemplified by unmanned vehicles with WSNs navigation. In: Proceedings of the 7th International Conference on Embedded and Multimedia Computing Technology and Service, Gwangju, Korea, pp. 269–275, September 2012
13. Wan, J., Yan, H., Liu, Q., Zhou, K., Lu, R., Li, D.: Enabling cyber-physical systems with machine-to-machine technologies. Int. J. Ad Hoc Ubiquitous Comput. **13**(3/4), 187–196 (2013)
14. Cucinotta, T., et al.: A real-time service-oriented architecture for industrial automation. IEEE Trans. Indus. Inf. **5**(3), 267–277 (2009)
15. Spiess, P., Karnouskos, S., Guinard, D., Savio, D., Baecker, O., Sa de Souza, L.M., Trifa, V.: SOA-based integration of the internet of things in enterprise services (2009). http://www.socrades.eu/Documents/objects/file1259604734.51
16. Troger, P., Rasche, A.: SOA meets robots - a service-based software infrastructure for remote laboratories. Int. J. Online Eng. **4**(2), 24–30 (2008)
17. Li, D., Li, F., Huang, X.: A model based integration framework for computer numerical control system development. Robot. Comput. Integr. Manuf. **26**(4), 333–343 (2010)
18. Li, F., Wan, J., Zhang, P., Li, D.: A multi-view integration language for cyber-physical robotic system. In: ICMLC 2013, pp. 387–392 (2013)

Research on the Link Quality Prediction Mechanism Based on ARIMA Model for Multi Person Cooperative Interaction

Shu Yao, Chong Chen, and Heng Zhang[(⊠)]

College of Computer and Information Science,
Southwest University of China, Chongqing 400715, China
dahaizhangheng@swu.edu.cn

Abstract. With the continuous enhancement of the hardware performance, the dynamic high speed sensing network in the small and medium size of the special industry is becoming more and more recognized and valued by the academic circles and enterprises, and in the process of multi - user interaction, the requirements on the quality of the link are also higher, but the high speed communication of the instability of the link quality and difficult to judge critical problem has not been solved. In this paper, we use the ARIMA model to predict the link quality of wireless multimedia sensor network, make the dynamic buffer and link switch in time. Finally through the experiment discovered after post-processing forecast network to meet the real-time dynamic environment average closing package rate, stability and robustness are significantly improved.

Keywords: Wireless multimedia sensor network · Cooperative interaction · Link quality · Trend forecast

1 Introduction

Wireless multimedia sensor network [1–3] is derived from the traditional sensor networks. It is a high speed wireless network which is dynamically self-organizing formed by a group of multi media sensor nodes with sensing, computing and communication capabilities, with the cooperative interaction among nodes, acquisition and processing of multimedia information, Ad hoc network [4–7] is a special kind of wireless mobile network, which can be used to communicate with the nodes outside the coverage area, and has great advantages in the field of military applications. Wireless multimedia sensor networks and Ad hoc networks are not new, with the continuous improvement of the hardware capabilities and application of innovative, in public security, military, anti-terrorism, industrial equipment and environmental monitoring and other fields based on this technology to build a small scale of the interactive work of the network reflects more demand and higher requirements.

1984 MIT Irene Greif and Paul Cashman from DEC, put forward the computer supported cooperative work. The main purpose of this is to establish collaborative work environment, improve information exchange between human beings. It's also can remove the obstacles in the time and space separated, improve the quality and efficiency

© ICST Institute for Computer Sciences, Social Informatics and Telecommunications Engineering 2016
J. Wan et al. (Eds.): Industrial IoT 2016, LNICST 173, pp. 231–239, 2016.
DOI: 10.1007/978-3-319-44350-8_23

of group work, make the group activities in the network environment become truly digital and agility [8–10]. With the rapid development of pervasive computing, wearable computing, the concept of the form and calculation of the computer is changing, and the extension of cooperation is also expanding.

2 BackGround

In a special environment, the environment is detected, the environment is matched with the real-time information, and the data is collected by the personnel or other equipment. It may happen that the action of high speed, static and slow moving, and the distance between nodes is constantly updated. At this point, the frequency of a single node in a certain range is very high, and the intensity of the sensor is also changed with the change of the distance. The transmission effect of the data packet is bad. Experiments in [11] show that the performance of the AOMDV protocol is declining very fast when the nodes move fast.

At present, the link quality is judged according to the link's remaining life cycle [12], value of RSSI [13–16], LQI [17, 18] or PRR [19]. Although the PRR value can directly reflect the current link state, it is needed to calculate the packet by a large number of samples, and waste a large amount of bandwidth and energy to obtain a large statistical base, and PRR is just a result of a certain time point that can only get the current or previous link quality, but link quality assessment of real-time in the application is very important. While LQI and RSSI can be used to reflect the quality of the link in the open environment without interference [20]. It is not desirable to use this method in our stochastic motion application environment.

In the [19], a average value of RSSI is used as a threshold for the link detection, and then the link is predicted in the future for a period of time. This is a critical point for the link quality of the above mentioned RSSI, which can be regarded as the normal transmission, when the data transmission is considered enough to affect the video quality, the link is switched to improve the transmission quality.

We will find that in the process of communication, data transmission sometimes get slowly when the broken. Although the data is transmitted, the effective transmission rate is very slow, giving a false impression of transmission. But the link is really connected state, routing can not switch because the link is disconnected, the transmission efficiency is very low. In [19] the experiment results show that the average RSSI is lower than 85 DBM, PRR values between 30 % ~ 95 % swings back and forth, link to present a very unstable state, while the average RSSI is greater than 85 DBM, PRR less volatile, link showed good stability. The link stability is poor when the transceiver signal is lower than a certain value, and the data communication is not stable at this time.

In previous studies usually just calculate the link quality after immediately to choose the best links to link to switch, the default link cannot achieve requirements before use. However, the change of the link quality in some application scenarios is a relatively slow process, calculate the quality link immediately after failure or does not meet the requirements of probability is very small. Generally, the two types of links are switched. Firstly, packet loss rate is high, but it is enough to guarantee the video quality is immediately switched; Secondly, packet loss rate has begun to affect the video quality before starting to calculate the link quality to switch, both of them make the high-speed data transmission process is interrupted, the frequent switching will cause the phenomenon of slow transmission of communication, if there are some very urgent interactive processes, it will result in very significant economic and personnel losses.

ARIMA model is a very famous time series forecasting method proposed by Jenkins and Boakes. The non-stationary time series is transformed to stationary time series, and the model is established by regression analysis of the lagged values of the variables and the present value of the random error term and the lag value of the random error term. The data is considered as a random sequence, and a certain mathematical model is used to describe the sequence, and the value of the past and the present value is predicted to be a trend.

In this paper, according to the problems of the above research, the paper puts forward the model based on the time series ARIMA model to analyze the PRR and gets a relatively long time period of the link change trend. According to the forecast trend, the link quality is calculated in advance, the preparatory action is carried out in advance, and some resources are used only in a certain time period.

When the transmission effect reaches a critical point, the transmission link is stable in a small fluctuation range, with a very small price to ensure the quality of data transmission, improve the network performance, solve the problem of the slow transmission of the critical communication phenomenon.

3 Model Construction and Algorithm

3.1 ARIMA Model

ARIMA(p,d,q) model is ARMA(p,q) model after d times difference, AR(p) is the Auto regressive model, p is a self regression, MA(q) is moving average model, q is moving average number; The Xt indicates the T period of the collection rate, and X has the characteristics of MA and AR, For example, the following is a ARMA (1,1) model, Θ is constant term.

$$X_t = \theta + \alpha_1 X_{t-1} + \beta_0 u_t + \beta_1 u_{t-1} \tag{1}$$

Constructing ARIMA(p,d,q) model, firstly should judge whether the time series is stationary sequence, according to the sequence of hash map, the autocorrelation function and partial autocorrelation function, the ADF unit root test variance and so on to carry out the identification of the sequence. As for the non-stationary time series, after d times difference:

$$\Delta X_t = X_t - X_{t-1} = X_t - LX_t = (1 - L)X_t \tag{2}$$

$$\Delta^2 X_t = (1 - L)^d X_t \tag{3}$$

After d times difference, we can get stationary time series $W_t = \Delta^d X_t$, ARMA modeling for W_t, you can get:

$$W_t = \emptyset_1 W_{t-1} + \emptyset_2 W_{t-2} + \cdots + \emptyset_p W_{t-p} + \delta + u_t + \theta_1 u_{t-1} + \theta_2 u_{t-2} + \cdots + \theta_q u_{t-q} \qquad (4)$$

The autocorrelation function of ACF ρ_k and the partial autocorrelation function of PACF φ_{kk} with model identification:

$$\rho_k = \frac{\sum_{t=1}^{N-K} X'_{t+k} X'_t}{N} \qquad (5)$$

$$\begin{cases} \varphi_{11} = \rho_1 \\ \varphi_{k+1,k+1} = \left(\rho_{k+1} - \sum_{j=1}^{k} \rho_{k+1} - j\varphi_{kj} \right) \left(1 - \sum_{j=1}^{k} \rho_j \varphi_{kj} \right)^{-1} \\ \varphi_{k+1,j} = \varphi_{kj} - \varphi_{k+1,k+1} \varphi_{k.k+1-j}, j = 1, 2, \cdots, k \end{cases} \qquad (6)$$

According to the above calculation results and according to Table 1 model identification principle, can be determined in accordance with the model.

Table 1. Model identification principle

Model	AR(p)	MA(q)	ARMA(p,q)
AF	exponential decay	finite length, truncation (q)	exponential decay
PAF	truncation (p steps)	exponential decay	exponential decay

In order to test the model satisfy the stationarity and invertibility of and need to estimate the parameters of the model, namely the following formula root meet outside the unit circle:

$$\varphi(B) = 1 - \sum_{j=1}^{p} \varphi_j B^j = 0 \quad \theta(B) = 1 - \sum_{j=1}^{p} \varphi \theta_j B^j = 0 \qquad (7)$$

It is also necessary to test the hypothesis of the model, and then the reliability prediction model can be obtained by the white noise test.

3.2 Energy Consumption Model

According to the prediction mechanism, the link quality is changed, and the link quality is calculated. The energy consumption model based on RSSI values is further constructed. Then, the residual energy of nodes is used as a factor to determine the quality of the link.

A node energy consumption model in wireless multimedia sensor networks is proposed in [17].

$$E_{new} = E_{old} - E_{consume} \qquad (8)$$

Node energy consumption $E_{consume}$ is mainly composed of data sending, receiving and forwarding of three parts. Nodes send data to consume energy:

$$E_{sent} = P_{sent}T_p = I_s VT_p \tag{9}$$

Forwarding data needs to consume energy:

$$E_{jorw} = (P_{sent} + P_{recv})T_p \tag{10}$$

Receiving data needs to consume energy:

$$E_{recv} = P_{recv}T_p = I_r VT_p \tag{11}$$

So the total energy consumption of the node:

$$E_{consume} = E_{send} + E_{recv} + (N-1)E_{jorw} \tag{12}$$

Among them, the P_{send} said the node transmit power, P_{recv} for receiving power, T_p for sending and receiving time, I_s for the transmission current, I_r for receiving current. RSSI as a node to receive signal strength indicator, and can be expressed as the node to send and receive power, The RSSI value of the current node is calculated as the energy consumption of the current node, and can be used to calculate the link quality in real time.

When the link quality is detected, the RSSI value of the node is detected as R. At the same time, calculate the node energy consumption as E, both of them as a link quality judgment standard. When the route switch is triggered, the impact of the hop count on the bandwidth is calculated with them. If the current node needs to transmit the video stream number more, increase the bandwidth of the weight, if only one way data need to be transmitted, it can increase the energy consumption and RSSI requirements. T_{max} can be selected when switching route,

$$T_{max} = \frac{1}{1+I^2} * (E_{min} + R_{max}) + \frac{I^2}{1+I^2} * \frac{K}{N} \tag{13}$$

K represents bandwidth, and N is the number of hop node and there are I video streams needs to be transmitted.

4 Test and Analysis

In this paper, ZYNQ-7000 is used as an experimental platform to extend the processing platform, using FPGA to combine software and hardware resources to achieve wireless multimedia sensor function. Two kinds of scenes, the building and outside garden, are tested and the data are tested. The nodes are distributed in the indoor and outdoor. In ad hoc networks, the time to complete the whole network interaction between nodes is required at most 7 frames, and the information synchronization is realized. So the data is collected by the 21 frame, and the random selection node is running, walking and waiting for the behavior to simulate a test environment.

After a period of time, the data collected is shown in Fig. 1.

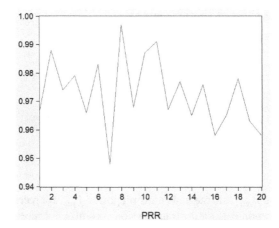

Fig. 1. PRR Time periodic value, the data fluctuate strongly, but still PRR almost 95 % or more

Obviously, the data prove that it's enough to ensure the transmission quality, ARIMA (2,2,1) model was determined by ARIMA model analysis, and the test residual sequence is qualified, we can consider the model can reflect the change trend of the actual collection rate.

Extending predicted space to get the model formula:

$$X_t = 0.960438 + 0.681309X_{t-1} + 0.251076X_{t-2} - 0.997366a_{t-1} \tag{14}$$

Extended space for trend prediction, as shown in Fig. 2.

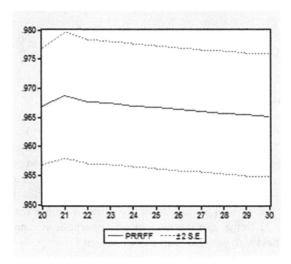

Fig. 2. Trend forecast results, the predicted value and the standard error of the graphics, in the future period of time the packet rate is a slow decline in the form of change.

It can be inferred that in the next period of time, the link transmission effect will gradually become worse. But it can be seen that the rate is still high, still maintained at more than 95 %, can continue to transmit data.

With the prediction of the trend, the PRR value indicates that the link quality will fall to the extent that the link is transmitted in the future.

The link quality can be calculated as the switch link to get the critical point, which indicates that the link transmission effect has been affected by the video quality.

Test the results, and there is a trend to judge and no trend to judge the results of the comparison of the results in Fig. 3;

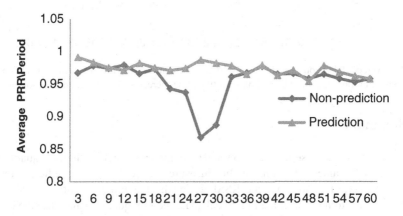

Fig. 3. Average packet rate change contrast, in the network without the use of forecasting mechanisms, the average rate of recovery has been a great decline after recovery

It can be guessed that the link quality is needed to be added after the critical point of the switch, and the link quality is needed to be switched. And the network's packet rate is predicted to have a certain growth trend in that time, which indicates that the time of the handover to the link quality is better.

At the same time, it can be seen in the time period of the test, which has a forecast mechanism of the network's packet rate stability in a very small range of fluctuations, and its transmission efficiency is able to meet the needs.

And for the no prediction mechanism network, the packet reception rate although in a certain period of time can also meet transmission needs, but link switching lead to instability, is likely to accept node suspended phenomenon. Compared the real time performance of data transmission (Fig. 4).

Without the use of predictive mechanisms in the network, it can be seen in the switching of the link time delay is very powerful, but also relatively large fluctuations. From the above evaluation index, the network with prediction mechanism, no matter on the stability, delay or robustness, its data transmission is better than that without the use of prediction mechanism.

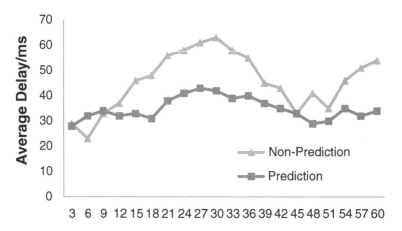

Fig. 4. Average delay contrast, the average delay of the network with a prediction mechanism is maintained in a range, although the link quality can be increased, but the change is not affected.

5 Summary

In this paper, we propose a time series ARIMA model to predict the link quality and select the PRR value as the link quality measure. The link quality prediction results show a downward trend, with the RSSI value, the remaining quality of nodes, the number of hops to the rest of the link to calculate the link quality, when the RSSI value reaches the critical point to switch the best link for data transmission. The results of simulation experiments show that the network packet rate, and the average end delay are stable in a small range, which is obviously superior to the network without using prediction mechanism. In real-time, robustness and stability more than before to meet the requirements of QOS, the basic solution to the data transmission is inefficient, resulting in slow transmission phenomenon.

In this paper, we only use the same link quality calculation method to compare the two networks, in the future testing, we also need to try to join other link calculation method and adding forecasting mechanism of the network to carry on the experiment, and compared with other prediction mechanism. In this paper, the single use of PRR value in the prediction of the link quality, in the follow-up work to increase the accuracy, may also need to refer to other factors, As for the critical point of the judgment is only a reference to the RSSI mean change, the current link quality changes can also be considered to join.

Acknowledgments. This article was supported by Fundamental Research Funds for the Central Universities (Grant numbers: SWU113066, XDJK2015C023), The ministry of education - Google co-operative professional comprehensive reform project-the base of Android application development technology (20710164), Special trade portable devices high-speed ad-hoc network protocol (41010815), and Southwest university school of computer and information science teaching team project.

References

1. Ma, H.-D., Tao, D.: Multimedia sensor network and its research progresses. J. Softw. **17**(9), 2013–2028 (2006). (in Chinese)
2. Monowar, M.M., Hassan, M.M., Bajaber, F., et al.: Thermal-aware multiconstrained intrabody QoS routing for wireless body area networks. Int. J. Distrib. Sens. Netw. **2014**, 1–14 (2014)
3. Chen, J.H., Song, J., Li, Y.: Research on application of WMSNs in ubiquitous learning environment. J. Adv. Mater. Res. **659**, 229–232 (2013)
4. Wang, H.-C., Woungang, I., Lin, J.-B., et al.: Revisiting relative neighborhood graph-based broadcasting algorithms for multimedia ad hoc wireless networks. J. Supercomputing **62**(1), 24–41 (2012)
5. Zhang, H., Zhang, Z., Zhang, F., Li, L., Wang, Y.: Optimized design of relay node placement for industrial wireless network. Int. J. Distriuted Sens. Netw. **2014**(2), 1–12 (2014)
6. Rodden, T.: A survey of CSCW systems. Internet Comput. **3**(3), 319–353 (1991)
7. Wang, Y.: Collision avoidance in muti-hop Ad hoc networks. In: Proceedings of the 10th IEEE International Symposium on Modeling Analysis and Simulation of Computer and Telecommunications Systems (MASCOTS 2002). IEEE, USA (2002)
8. Savetz, K., Randall, N., Lepage, Y.: MBONE: Multicasting Tomorrow's Internet. Hungry Minds Inc., US (1996)
9. Shi, M.-L.: CSCW: computer supported cooperative work system. Chin. J. Commun. **1995**(1), 55–61 (1995). (in Chinese)
10. Shi, M.-L., Xiang, Y.: Key techniques in CSCW research. Chin. Acad. J. **3**(11), 1389–1392 (1997). (in Chinese)
11. Marinal, M.K., Das, S.R.: Ad hoc on-demand multipath distance vector routing. J. Wirel. Commun. Mob. Comput. **6**, 969–988 (2006)
12. Wang, Q.-W., Shi, H.-S.: Ad Hoc QoS network link quality multipath on-demand routing protocol. Comput. Eng. Appl. **46**(29), 29–32 (2010). (in Chinese)
13. Ndzi, D.L., Harun, A., Ramli, F.M., Kamarudin, M.L., Zakaria, A., Md. Shakaff, A.Y., Jaafar, M.N., Zhou, S., Farook, R.S.: Wireless sensor network coverage measurement and planning in mixed crop farming. Comput. Electron. Agric. **105**, 83–94 (2014)
14. Sun, P.-G., Zhao, H., Pu, M., Zhang, X.-Y.: Assessment of communication link in wireless sensor networks. J. Northeast. Univ. (NATURAL SCIENCE EDITION) **29**(4), 500–503 (2008). (in Chinese)
15. Srinivasan, K., Levis, P.: RSSI is under appreciated. In: Proceedings of the 3rd Workshop on Embedded Networked Sensors (EmNets), pp. 1–5 (2006)
16. Sawant, R.P.: Wireless Sensor Network Testbed: Measurement and Analysis. The University of Texas at Arlington, Texas (2007)
17. Alec, W., David, C.: Evaluation of Efficient Link Reliability Estimators for Low-Power Wireless Networks, pp. 1–20. UC Berkeley, LosAngeles (2002)
18. Wang, Y., Martonosi, M., Peh, L.S.: A new scheme on link quality prediction and its applications to metric based routing. In: Proceedings of the 3rd International Conference on Embedded Networked Sensor Systems, pp. 288–289. ACM Press, SanDiego (2005)
19. Huang, T.-P., Li, D., Zhang, Z.-L., Cui, L.: An adaptive link quality estimation method for sudden link perception. J. Comput. Res. Dev. **47**(Suppl.), 168–174 (2010). (in Chinese)
20. Cheng, D.-W., Zhang, X.-Y., Zhao, H.: Study routing metrics based on EWMA for wireless sensor network. Sens. Technol. **21**(1), 65–69 (2008)

Design of Multi-mode GNSS Vehicle Navigation System

Zhijie Li[1], Jianqi Liu[1(✉)], Yanlin Zhang[1], and Bi Zeng[2]

[1] School of Information Engineering, Guangdong Mechanical and Electrical College,
Guangzhou, China
{5294968,350054049}@qq.com, liujianqi@ieee.org
[2] Guangdong University of Technology, Guangzhou, China
zb9215@gdut.edu.cn

Abstract. As the Beidou System (BDS) and Galilei Positioning system gradually improve, people intend to make use of more satellites to offer location service for vehicle and other mobile terminal. A kind of multi-mode vehicular positioning system is proposed, the whole framework can be divided into three parts as XN647-8 positioning module, signal processing module and the antenna. XN647-8 is a high-performance, low-cost single-chip GNSS receiver module. Its signal processing module includes a high gain, low noise amplifier chip XN114 and surface acoustic wave (SAW) filter circuit. Through a oscilloscope and analyzer, BDS/GPS antenna is measured and calibrated. So multimode receiver can actually receive satellite signal strength indication close to the theoretical value. The experimental results show that the receiver have a so good performance that it can achieve real-time location tracking well.

Keywords: GNSS · Vehicle navigation · Positioning and tracking · Multi-mode satellite positioning

1 Introduction

The accuracy of satellite positioning system has been increased substantially after a long term development. However, the more accurate positioning requirements is proposed when the GNSS-based vehicular positioning system is used in new application of the Internet of Vehicles (IoT) [1–3]. The positioning accuracy and the user experience should be improved substantially. Improving positioning accuracy not only is target for mainstream navigation services vendor, who make arduous efforts abidingly, but also is the important technical support which will be extended to a variety of consumer electronics products [4, 5]. But now it is also difficult to satisfy the full range of application requirements depending on a single satellite positioning system technology [6].

DBS is growing very rapidly. For example, a new generation Beidou satellite is launched successfully which is a landmark of the building of 2020 Global Network. DBS entered into the "3.0 times". Global networking will be realized gradually. The same change is happening on Galileo Navigation Satellite System. This means that the traditional GPS is transferring into the GNSS [7–10]. The availability, interference, etc. of GNSS are better than GPS's. Its multi-mode receiver has higher cost performance than the single GPS receiver. Multimode-compatible receiver is the inevitable direction

© ICST Institute for Computer Sciences, Social Informatics and Telecommunications Engineering 2016
J. Wan et al. (Eds.): Industrial IoT 2016, LNICST 173, pp. 240–246, 2016.
DOI: 10.1007/978-3-319-44350-8_24

of future development. This paper will describe the building of GNSS positioning system based on multi-mode chip XN647-8.

2 BDS/GPS Multimode Satellite Receiver Framework

GNSS positioning system is divided into three parts, XN647-8 positioning module, signal processing modules (low-noise amplifier module with SAW filter modules) and antennas. The system framework is shown in Fig. 1.

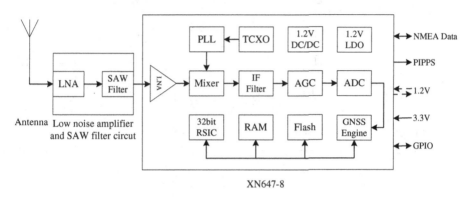

XN647-8

Fig. 1. The block diagram of GNSS positioning system

XN647-8 is a high-performance, low-cost single-chip GNSS satellite navigation receiver module with low power consumption, high sensitivity, small size and other characteristics. It can be received in parallel multimode satellite navigation signals. XN647-8 highly integrated system solutions. Its components include Integrated RF receiver unit, Baseband processing unit, a temperature-compensated crystal oscillator (TCXO/0.5 ppm), low dropout linear regulator (LDO) and some passive components (resistors, capacitors, inductance). Only a few external resources are used to integrate the mobile terminals and handheld devices.

XN647-8 has a strong ability to capture weak signals and a short first position time. Correlating signal parameter search engine base with internal dedicated to quickly capture the available satellites, and can receive a weak signal. Even in harsh environments, advanced tracking engine is also capable of weak signal tracking and positioning.

Its main features: ① small multimode GNSS receiver module, support for multiple satellite positioning system; ② high sensitivity, cold start sensitivity: −148 dBm; tracking sensitivity: −165 dBm; ③ fast start-up time, cold start requires 29 s, hot start only need 1 s; ④ positioning accuracy of 2.5 m; ⑤ easy programming, support for external SPI flash memory data record. According to the instructions and circuit requirements XN647-8 chip, design positioning module circuit diagram is shown in Fig. 2.

242   Z. Li et al.

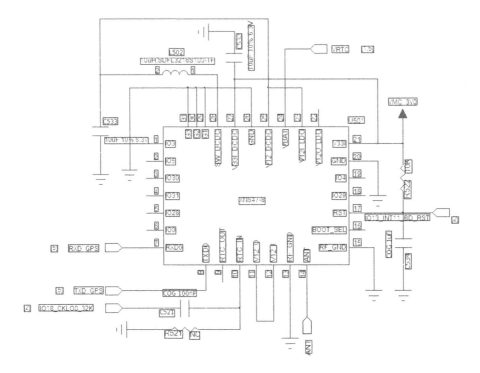

Fig. 2. Low noise amplifier and surface acoustic wave filter circuit schematic diagram

3 Low Noise Amplifier and Surface Acoustic Wave Filter Circuit

XN114 is a chip which is usually be applied to a low noise amplifier receiver. It is designed with BiCMOS technology, with high gain and low noise characteristics, the conditions under 18 dB of gain, noise figure reached 0.8 dB. The chip is applied to the receiver front end, which can effectively improve the receiving sensitivity of the receiver and expand the scope of application of the receiver. XN114 can work in +1.8 V~+ 3.3 V supply voltage while current consumption is only 5.5 mA. In shutdown mode the current

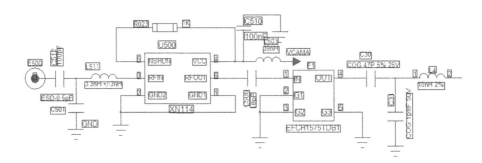

Fig. 3. Low noise amplifier and surface acoustic wave filter circuit schematic diagram

consumption is less than 10μA. It is packaged with 6-pin LGA form, which size is 1.5 mm × 1.0 mm × 0.75 mm. Recommended operating conditions: power supply voltage1.8 V~3.3 V, Operating temperature −40 °C~85 °C. According to The circuit chip demand and user guidance, we designed low noise amplifier and sound table filter circuit. The schematic diagram shown in Fig. 3.

4   Measurement and Calibration of BDS/GPS Antenna

BDS/GPS antenna is defined as a kind of antennas which is used in the satellite positioning system for receiving the positioning or navigation signals used. BDS antennas work at the center frequency of 1561.098 MHz while GPS L1 works at a center frequency of 1575.42 MHz. Signal power is generally about −166 DBM, belongs weak signal. BD/GPS antenna has four important parameters: gain, voltage standing wave ratio (VSWR), noise figure (Noise figure), the axial ratio.

Signal generally about −166 DBM, belongs to a weak signal. BDS/GPS antenna has four important parameters: Gain, standing wave ratio, noise figure, the axial ratio.

BDS/GPS antenna performance is measured by a oscilloscope and analyzer (HP 8752A). Measurement procedure is as follows:

1. Use calibrations to correct network analyzer by single port mode.
2. Connecting the test antenna to the network analyzer.
3. Measuring the reflection coefficient (S11), return loss and other relevant parameters.

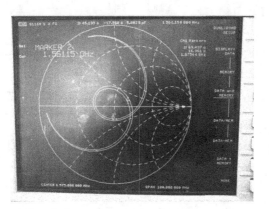

Fig. 4. The smith chart of BDS/GPS antenna

BDS/GPS antenna sizes are 25 mm × 25 mm × 4 mm. According to measurement procedures relevant graphical data is obtain obtained. Figure 4 is a Smith chart of the antenna while Fig. 5 is the VSWR diagram.

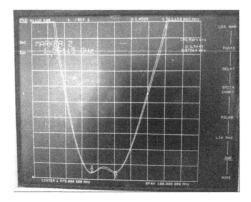

Fig. 5. Standing wave chart of BDS/GPS antenna

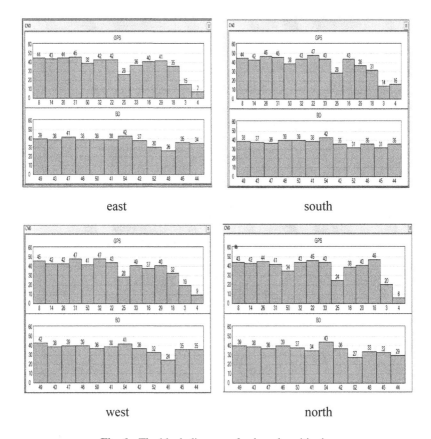

Fig. 6. The block diagram of onboard positioning

Figure 6 give the machine actually received satellite signal strength value when it is put different orientations at outdoor occasions: the east, south, west and north. Figure 5 shows good BDS/GPS antenna performance. Actually received satellite signal strength values of the machine is close to the theoretical value.

5　GNSS Positioning Module Satellite Capture Test

First needs of GNSS receiver is to capture the satellite signal. The minimum signal strength required to complete capture is named capturing sensitivity. After the capture, the satellite signal tracking sensitivity is very important parameter. It is defined as the minimum signal strength to maintain the continuous tracking.

Performance positioning module for positioning the circuit to be tested. Positioning chips collected by BDS/GPS antenna satellite system signal. After amplification filter circuit, the satellite signal is sent to XN647-8 Monolithic GNSS satellite navigation receiver module. Then the data changed to output by the ADC and internal operations according to NMEA protocol.

Overall performance test results of GNSS positioning system shown in Figs. 5, 6 and 7. The figure includes the first positioning time, TTFF different protocol selection, RTC, receiver sensitivity, and satellite location data. As can be seen, the vehicle

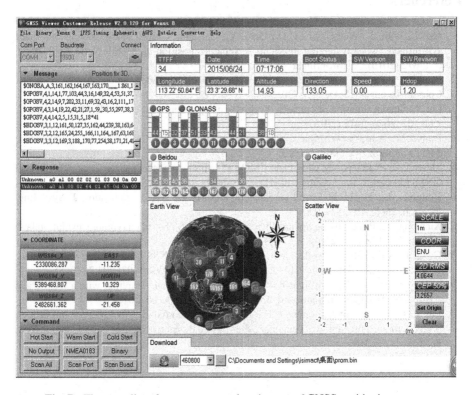

Fig. 7. The overall performance comprehensive test of GNSS positioning system

networking intelligent terminal receives 9 GPS satellite signals and 6 BDS signals. The specific satellite position is also can be identified. Vehicle networking intelligent terminal device owns good positioning performance that can achieve real-time location tracking.

6   Conclusion

This design improved GNSS-based vehicle positioning system. System hardware framework and integrated control terminal are designed to improve the integration of the system. Overall performance test results showed that: vehicle networking intelligent terminal can successfully receive signals from multiple GPS satellites and BD satellites, can be a good vehicle to achieve real-time location tracking.

Acknowledgments.   The authors would like to thank Guangdong Province Special Project of Industry-University-Institute Cooperation (No. 2014B090904080), 2013 Guangdong Province University High-level Personnel Project (Project Name: Energy-saving building intelligent management system key technologies research and development) and the Project of Guangdong Mechanical & Electrical College (No. YJKJ2015-2) for their support in this research.

References

1. Liu, J., Wan, J., Wang, Q., Li, D., Qiao, Y., Cai, H.: A novel energy-saving one-sided synchronous two-way ranging algorithm for vehicular positioning. Mobile Netw. Appl. **20**(5), 661–672 (2015). ACM/Springer
2. Schmidt, G.T.: Navigation sensors and systems in GNSS degraded and denied environments. Chin. J. Aeronaut. **28**(1), 1–10 (2015)
3. Liu, J.: State estimation of connected vehicles using a nonlinear ensemble filter. J. Cent. S. Univ. **22**, 2046–2415 (2015)
4. Binbin, W.: A low power high gain-controlled LNA mixer for GNSS receivers. J. Semicond. **34**(11), 115002:1–115002:7 (2013)
5. Shimao, X.: A low power wide-band CMOS PLL frequency synthesizer for portable hybrid GNSS receiver. Chin. J. Semicond. **31**(3), 035004:1–035004:5 (2010)
6. Hongmei, L.: Novel dual-band antenna for multi-mode GNSS applicatios. J. Syst. Eng. Electron. **26**(1), 19–25 (2015)
7. Xianqing, Y.: Route strategy of satellite network in GNSS based on typology evolution law. J. Syst. Eng. Electron. **25**(4), 596–608 (2014)
8. Gang, J.: A digitally controlled AGC loop circuitry for GNSS receiver chip with a binary weighted accurate DB-linear PGA. J. Semicond. **36**(3), 035004:1–035004:7 (2015)
9. Xiangyang, Y.: A binary-weighted 64-DB programmable gain amplifier with a DCOC and AB-class buffer. Chin. J. Semicond. **33**(2), 025003:1–025003:6 (2012)
10. Liu, J., Wan, J., Wang, Q., Zeng, B., Fang, S.: A time-recordable cross-layer communication protocol for the positioning of vehicular cyber-physical systems. Future Gener. Comput. Syst. **56**, 438–448 (2016)

Intelligent Storage System Architecture Research Based on the Internet of Things

Li Liu[✉] and Kaifeng Geng

Nanyang Institute of Technology, Nanyang 473000, Henan, China
liuli861202@126.com

Abstract. With the development of e-commerce, Internet of things (IOT) and computer network technology, logistics has an increasingly important status is daily life. However, at the same time of providing people with convenience, e-commerce puts forward higher challenge to the storage link [1]. Proceeding from the problems such as the lagging logistics work caused by uneven flow of information, this paper investigated the impact of IOT on the warehousing industry, proposed the IOT based intelligent warehouse system model, which can promote the business flow by warehouse information flow and achieve the purpose of improving the upper layer business operation.

Keywords: Internet of Things · RFID · Intelligent storage · Sensing technology

1 Introduction

In recent years, the explosive development of e-commerce has promoted the rapid rise of the logistics industry. However, the surge of online shopping orders also exposed the weaknesses of the industry. As a result, the original storage and preservation function has been far from satisfying the needs of the industry. The prominent blasting warehouse, delayed orders, goods out of stock, damaged goods and other phenomenon, does not only bring a lot of inconvenience to customers, but also cause huge negative impact on suppliers and e-commerce platform service providers. Thus, the importance of storage in logistics is increasingly prominent. As the link connecting manufacturers, platform service providers and consumers, it can help overcome the spatial and temporal differences between the two parties, support the enterprise logistics decision, while it can also achieve the purpose of improving customer service level and reducing logistics cost.

The warehousing link has a lot of work including loading and unloading, handling, stacking, statistics and information transfer, which is mostly done by manual operation at present. China's warehousing industry features large holdings, relatively low management level, and imbalanced warehousing internal technical development. In spite of the substantial improvement of the storage infrastructure and the emergence of many stereoscopic warehouses, the warehousing technology application level has made some progress, but still cannot meet the need.

Relying on the promotion of national policy and considering the restriction and impact brought by the low warehousing efficiency, the development of IOT technology based intelligent warehouse system will better solve the current warehouse inventory,

© ICST Institute for Computer Sciences, Social Informatics and Telecommunications Engineering 2016
J. Wan et al. (Eds.): Industrial IoT 2016, LNICST 173, pp. 247–256, 2016.
DOI: 10.1007/978-3-319-44350-8_25

meet the needs of warehousing business in e-commerce environment to significantly improve the service level of the stock sectors [2].

2   The Impact of IOT Technology on Warehouse Management

2.1   Optimize Warehouse Operation Process and Improve the Level of Warehousing Services

The study of IOT technology based intelligent storage system is conductive to improve the low efficiency in storage, inventory, environmental monitoring and other links, improve the information transmission efficiency and realize the real-time share in the inventory, environment, output and input of warehouse [3]. IOT technology has optimized the traditional storage process, achieve real-time, accuracy and high efficiency from information acquisition, processing to the application of the upper business information, give full play to the role of warehousing link deployment and improve the level of storage service.

2.2   Improve the Level of Warehousing Intellectualization and Informationization

The transformation from manual operation, mechanization operation to intellectualization reflects the accelerating rotation speed of the storage corporate entity flow. In the e-commerce environment, the promotion of website operation, goods launching and shopping settlement become extraordinarily directly and efficiency; the integration and distribution of cross-enterprise and cross-regional goods is more prosperous, which continues to drive the development of virtual economy and real economy and promotes the seamless delivery of business flow and logistics [4]. Urged by the powerful impetus of economic interests, warehouse management will develop towards the direction of being more intelligent to achieve the purpose of on-demand integration and distribution, handle the processing and information management tens of thousands of orders every day, so as to complete the docking of business flow and logistics. Under the environment of IOT technology, to provide every customer with the best quality service, warehousing enterprises continues to improve the information-based degree to implement the intelligent level of warehouse management.

2.3   Realize the Information Synchronization and Sharing Between Enterprises in the Supply Chain

To warehousing sector, the geographic separation has become an important part of supply chain business management. The centralized management mode exists in name only under the environment of the current land policy [5]. Each enterprise should consider its own operating costs, especially in the electronic commerce environment. In the operation process of small profit mode, scattered remote management will better continue the integration thinking of supply chain management to achieve the supervision, management and control of remote warehouse sector. The strengthening of the information sharing between the warehousing link and upstream and downstream link in the supply chain is conductive to the mutual support and convergence between the businesses in the whole supply chain, also effectively solves the problem of information asymmetry in the supply chain management integration, successfully alleviates the difficult problems in land requisition, improve the overall efficiency of supply chain management and the informatization level of warehouse sector management.

3 Study of Key IOT Technology

The application of key IOT technology provides a good technical foundation for building an intelligent warehouse. The integrated application of IOT technology in the warehouse industry ensures the transfer efficiency and accuracy of key data information flow in the intelligent warehousing, making a great contribution on solving the problem studied by this paper.

3.1 RFID Technology

RFID is the abbreviation of radio frequency identification, which is a non-contact automatic identification technology. It can automatically identify the target via radiofrequency signal and obtain the related data in the object. By transmitting into the computer system, these data can complete the track, identification, storage, sensing and other processes [6]. The working principle of RFID system is shown in Fig. 1. When the label enters the magnetic field induction area and emits the carried information to the reader, the reader will obtain the information through sending a certain radio frequency via antenna and transmit the information to host computer for processing.

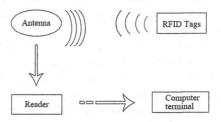

Fig. 1. RFID working principle diagram

The efficiency comparison of three input mode such as RFID, bar code technology and artificial entry is shown in Table 1 [7].

As the above characteristics of RFID, It can provide the most direct individual recognition to realize the link of everything in the world, which is seen as the most important one of the key technologies of Internet of things. The same technology can be used in modern logistics system, which can not only improve the ability of warehouse loading and unloading, inventory, product tracking in logistics enterprise, but also can

Table 1. Entry rate comparison

Entry requirements	1 piece	10 pieces	100 pieces	1000 pieces
Artificial entry	10 s	100 s	1000 s	2 h 47 min
Bar code technology	2 s	20 s	200 s	33 min
RFID	0.1 s	1 s	10 s	1 min 40 s

provide better service for the customer, and lay the foundation for realizing automatic and intelligent logistics system [8].

3.2   ZigBee Technology

ZigBee technology is a two-way wireless communication technology featuring close range, low power consumption, low rate and low cost, which is mainly used in the data transmission between various electronic devices characterized by close range, low power consumption, low transmission rate and typical periodic data, intermittent data and low response time data transmission [9]. Except the above characteristics, ZigBee technology can implement object positioning indirectly, which make it rapidly applied to storage environment management field of logistics management link.

The data frame is fully reflected in ZigBee, each wireless data frame includes a large number of wireless packages, including the information of a lot of time, address, order and synchronization. The real data information only accounts for a small part, becoming the key for ZigBee to realize network organization, high reliability and lower power transmission.

3.3   Sensor and Detection Technology

In the national standard GB/T 7665-2005, the sensor is defined as follows: devices can feel being measures and convert into the available output signal according to certain rules, usually consisting of sensitive components and conversion components [10]. Sensor is a kind of testing device, which can feel the measured information and convert the information into electrical signal or other information outputs by certain rules to meet the requirements of information transmission, processing, display and control, and is the first step to achieve automatic monitoring and automatic control. With the development of technology, the sensor is gradually realizing miniaturization, intelligentization and networking.

4   IOT Technology Based Intelligent Storage System Model Architecture

Storage system is a very complicated system, requiring unified management of the hardware, software and goods in the warehouse to achieve the intelligentization of warehouse management. In order to realize the synchronization of data information flow and the real flow in warehouse management process in time and space, intelligent storage system architecture needs to pay attention to the various aspects of the operation information such as upload, processing, parsing and encapsulation and so on, so as to maintain the integrity of information and speed up upload and processing capacity of data, so that the data can quickly build associated with the upper business information, and provide effective data to support the upper business, which will speed up the efficiency of upper business. Intelligent warehouse system prototype is extracted according to the research of typical business function and carry out the study of information flow in the late intelligent warehousing based on the prototype architecture, so as to improve the level of information technology and accelerate the warehouse operation efficiency [11]. The intelligent storage system model is as shown in Fig. 2.

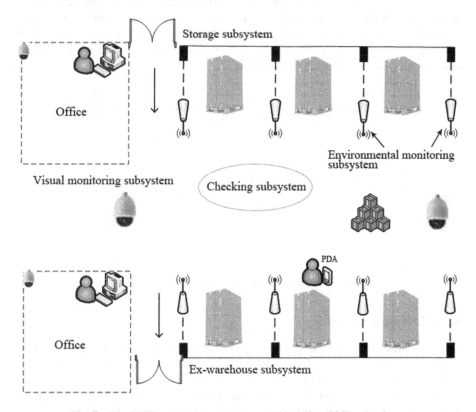

Fig. 2. The intelligent storage system model based on IOT technology

It can be seen from the picture above, the intelligent storage prototype system mainly includes the in-out stock subsystem, Checking system, environment monitoring subsystem and visual monitoring subsystem; orderly connection between systems is indispensable from the fast processing and transfer of information flow to support and meet the demand of warehouse system.

4.1   Storage Subsystem

Storage subsystem is mainly realized by RFID and serial communication technology. Hardware facilities mainly include RFID tag, RFID antenna, upper computer, portal access to ensure the goods data access and rapid transfer of information. In the goods circulation process, other information achieved real-time transmission via RFID tag, reader and main control system, so as to complete the transparent management of goods in the warehouse.

In the process of goods warehousing, goods and circulation unit is granted a sole ID respectively to complete the binding of goods and circulation unit with RFID tag [12]. It is used as the information carrier of goods and the circulation unit in the entire circulation to ensure the goods information track throughout the warehouse management. When the warehouse administrator uses PDA devices to complete the goods shelves scanning, PDA will transmit the scanning information to the upper machine for saving. Business function flow chart is shown in Fig. 3:

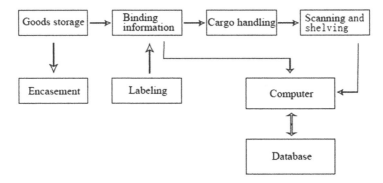

Fig. 3. Storage subsystem business flow diagram

4.2   Checking Subsystem

Checking subsystem refers to that when goods arrive at the designated position, bind the goods with the corresponding position information. In the process of goods storage management, each move and position change requires to bind the goods and position and rapidly get the goods RFID tag information via smart shelves, PDA scanner, etc. In this way, the accurate control of goods position can be achieved. At the same time, by the scanning of RFID, you can achieve real-time grasp of goods quantity information and dynamic changes by the scanning of RFID antenna and avoid the economic loss caused by inadequate inventory or excessive inventory. It cannot only achieve the accurate positioning and rapid goods inventory [13].

The entire flow of information transfer achieved the purpose of real-time, efficient, accurate, smooth and automatic data update, meet the needs of high-level vertical shelves, realized the consistency of physical flow and information flow, solved the synchronization problem of goods flow and information flow on shelves in time, improved the operational efficiency of the whole process.

4.3   Environmental Monitoring Subsystem

In warehouse storage, the goods environmental information collection is extremely important. The change of goods external environment will affect the changes of the good itself, sometimes will even lead to the reduction or scrapping of goods service life. Environmental monitoring subsystem is to use WSN/ZigBee technology to deploy each terminal node in every shelf of the warehouse, and collect the current shelf environment information (including temperature, humidity, smoke, etc.), and get the terminal node information to upload to the upper machine. Then the upper machine will judge whether the monitored shelves are in a secure environment according to the upper and lower threshold value. In the case of above or below the threshold, the system will automatically trigger a warning system to remind warehouse management personnel for inspection, realizing the intelligent management of warehouse goods external environment. The environment node deployment diagram is as shown in Fig. 4:

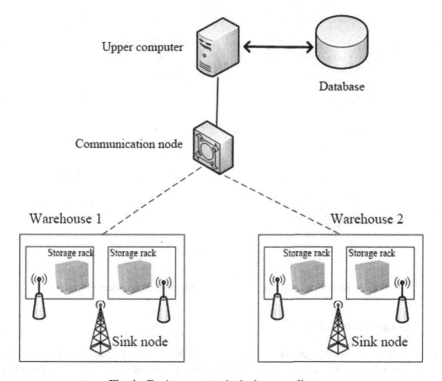

Fig. 4.   Environment node deployment diagram

4.4   Visual Monitoring Subsystem

Visual monitoring subsystem aims to achieve simplified, intellectualized and visualized warehouse internal management. Its hardware facilities mainly include a webcam, LED display, speech broadcast audio equipment, etc. By deploying the web camera in the warehouse, the entire warehouse operation can be accurately monitored to achieve sound

254   L. Liu and K. Geng

supervision and regulation of internal warehouse personnel operation flow (such as unsafe operation habit, goods lost verification, etc.) [14]. LED display can make real-time display of environmental monitoring information for internal staff viewing. At the same time, it can be used as the board of internal warehouse operation flow to display the next operation task, improve the execution efficiency of operators, making it become the carrier of warehouse internal information to efficiently accelerate the transfer speed. Voice broadcast can remind the warehouse management personnel of the internal warehouse monitoring information in the form of voice to complete the automatic push function of environmental information. This information can be transmitted to the local monitoring end via cable network and can also be transmitted to the remote monitoring terminal via wireless network to facilitate the remote monitoring of management sector. Remote monitoring subsystem equipment deployment diagram is as shown in Fig. 5:

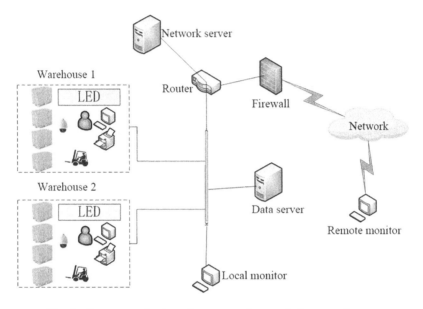

Fig. 5. Remote monitoring subsystem equipment deployment diagram

4.5 Ex-warehouse Subsystem

Ex-warehouse subsystem greatly improves the efficiency and accuracy of goods ex-warehouse, which cannot only achieve the real-time sharing of ex-warehouse to facilitate real-time storage materials, alleviate the problem of excessive or insufficient inventory caused by lagging ex-warehouse information; at the same time, it also provides information security to the warehousing backend transportation monitoring in the supply chain integration.

Ex-warehouse subsystem hardware equipment includes entrance guard (RFID read-write antenna, upper computer and serial line), the portal channel makes the inventory of ex-warehouse goods via RFID reader-write device. During ex-warehouse, the forklift or other transport tools will transport the goods through the ex-warehouse channel.

Because of the different pallet code and goods code, you can access to the tag information by RFID antenna and upload to the workbench terminal by serial communication. And then make comparison between the acquired information with the ex-warehouse information in the upper machine. If no error is found, the goods will smoothly pass the portal guard and complete the ex-warehouse business.

There is a strong dependency between subsystems. The smooth link in business largely depends on the completion status of the upstream subsystem, of which the main cohesion is the transfer of information flow [15]. To a certain extent, the efficiency of information flow transfer determines the efficiency of related business of each functional modules; the subsystems complete the purpose of passive launching to active warehousing business of information flow, achieve the purpose of business information as the driving force to the physical movement of goods in the warehouse from the subsystem storage input to storage output.

5   Conclusion

The application of IOT technology is gradually popularized in various industries and the research on the warehousing link informationization in the logistics industry is the focus of future research. This paper focuses on the discussion of the related issues of lagging information flow in the warehouse, which is combined with the application of many hardware facilities and IOT related technologies such as RFID, ZigBee technology. It puts forward the IOT technology based intelligent warehouse system model, makes in-depth analysis of each module and function in the model, aiming to improve the informationization level of warehousing system and storage efficiency.

References

1. Ye, G., Hua, C., Yang, M.: Research and design of visual intelligent warehouse information management system. Software **2**, 64–66 (2012). (in Chinese)
2. Zhang, R., Li, Y.: IOT technology based warehouse management system research. Zhengzhou Univ. **6**, 18–36 (2009). (in Chinese)
3. Yang, W.: Streamlined warehouse management counter measures. J. Logist. Technol. **7**, 360–362 (2012). (in Chinese)
4. Chow, H.K.H., Choy, K.L., Lee, W.B., et al.: Design of a RFID case-based resource management system for warehouse operations. Expert Syst. Appl. **30**(4), 561–576 (2006). (in Chinese)
5. Xiao, W.: Modern Logistics Intelligent Warehouse System Security Monitoring Technology and Simulation Implementation. Wuhan University of Technology (2006). (in Chinese)
6. Wan, Y.: Research on RFID Technology Based Logistics Storage Standard System. South China University of Technology, Guangzhou (2011). (in Chinese)
7. Yan, L., Zhang, Y., Yang, L.T., Ning, H. (eds.): The Internet Of Things: From RFID to the Next-Generation Pervasive Networked Systems. CRC Press, Boca Raton (2008). (in Chinese)
8. Baoyun, W.: Review on internet of things. J. Electron. Measur. Instrum. **23**(12), 1–7 (2009). (in Chinese)
9. Zigbee technology. http://baike.sogou.com/v7756286.htm. (in Chinese)

10. Zhu, Z.: Progress and trend of sensor network and internet of things. Microcomput. Appl. **1**(26), 1–3 (2010). (in Chinese)
11. Zhang, B., Liu, X.: WSN and RFID based intelligent warehouse management system design. J. Commun. Univ. China Nat. Sci. **9**, 38–40 (2009). (in Chinese)
12. Xue, Y., Liu, H.: Intelligent storage and retrieval systems based on RFID and vision in automated warehouse. J. Netw. **7**(2), 365–369 (2012). (in Chinese)
13. Liu, G., Yu, W., Liu, Y.: Resource management with RFID technology in automatic warehouse system. In: IEEE/RSJ International Conference on Intelligent Robots and Systems, pp. 3706–3711. IEEE (2006). (in Chinese)
14. Zheng, L.R., Nejad, M.B., Zou, Z., et al.: Future RFID and wireless sensors for ubiquitous intelligence. In: NORCHIP 2008, pp. 142–149. IEEE (2008). (in Chinese)
15. Hai, Z.: Research on IOT Based Warehouse Management System and the Middleware. Nanjing University of Posts and Telecommunications, Nanjing (2011). (in Chinese)

Design of Remote Industrial Control System Based on STM32

Rongfu Chen, Yanlin Zhang$^{(\boxtimes)}$, and Jianqi Liu

School of Information Engineering, Guangdong Mechanic and Electrical College,
Guangzhou 510550, China
{343656182,350054049}@qq.com, liujianqi@ieee.org

Abstract. Pushed by "Internet +" and Industry 4.0, the production mode of traditional industry will be changed by technological innovation. Since the core is intelligent manufacturing, the manufacturing industry will be intelligentized and internetized. Nowadays, however, most of the manufacturing facility cannot meet the requirement for intelligent manufacturing. Therefore, it's necessary to design a remote industrial control system, having STM32F407 be the master CPU since its operating frequency can reach 168 MHz and STM32F103 be the node CPU since its cost performance is quite impressive. The master connects to the node with RS485. The master can also be the gateway, responsible for data exchange between intranet and server, while the node can detect sensors and control actuators. This system helps to realize intelligentization and internetization, meanwhile, administrators can monitor and control the system through computer and mobile terminals APP.

Keywords: Remote monitoring · STM32 · Cloud service · MODBUS

1   Introduction

Currently, old-fashioned production control system is still widely adopted by most manufacturers. In this kind of production control system, each process node is independent, meanwhile, one worker is needed to monitor and manipulate one process node or several, which causes the waste of resource, the discrepancy between products due to the competence of each worker, and thereby inefficiency in the enterprise. With the promotion of Industry 4.0 [1], smart plant is popularized, therefore, that old-fashioned control system will become obsolete.

To solve the problem mentioned above, for better stabilization and manipulation, one remote industrial control system is designed with 3 layers: node perception, master control gateway and cloud service. Node perception layer is responsible for collecting field data, processing simple data, packing data and sending to master control gateway; the gateway will encrypt the received data and send to cloud server by package; the server can complete data analysis, obtain the optimal control parameters, feedback to the control system and manipulate the system. This remote industrial control system helps to automate the production system, optimize the allocation of resources, achieve uniform quality [2], and improve the efficiency during production process.

© ICST Institute for Computer Sciences, Social Informatics and Telecommunications Engineering 2016
J. Wan et al. (Eds.): Industrial IoT 2016, LNICST 173, pp. 257–267, 2016.
DOI: 10.1007/978-3-319-44350-8_26

1.1   Diagram of System Composition

The node perception layer of this remote industrial control system is composed of various nodes and each node can complete one process in the industrial production. Also, the node perception is capable of actuator control, monitoring the parameter of a certain procedure according to the received control signal. As a transfer station of the whole system, the master control gateway is in charge of collecting the uploaded data of each node, completing encryption & protocol conversion, sending that information to the cloud server, receiving the control signal from the cloud server, conducting decryption & protocol conversion [3], and modulating the parameter of relevant nodes. User administration terminal is composed of PC or mobile terminal. The system diagram is shown in Fig. 1.

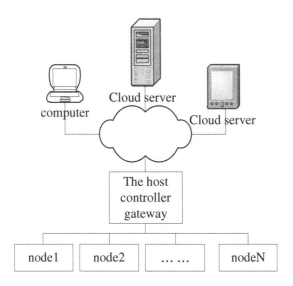

Fig. 1. System composition block diagram

1.2   Data Transmission

Data transmission is divided into two parts: LAN and WAN. LAN refers to master control gateway and each node, while WAN is the cloud server. PC and mobile terminal can choose LAN or WAN according to its network access point.

Under the MOBUS protocol, the main controller connects with the nodes through two-wire line 485. The main controller is the host mode and the nodes are slave mode. The host can read the slave by roll poling [4]. Communication format is as follows:

(1)   The default format for communication is 8, N, 1 The default baud rate is 19200 bps

(2) protocol is MODBUS RTU [5], Register operation is shown in Table 1:

Table 1. The register list

SN	address	Instructions
1	0000H	read-only, model, Value A001
2	0001H	read-only, Device type B002, Representative is a node
3	0002H	Read, write, node address
4	0038H	read-only, Flow of A size
5	0039H	read-only, Flow of B size
6	003AH	read-only, Reaction zone temperature
7	003BH	Read, write, Control valve "A"
8	003CH	Read, write, Control valve "B"
9	003DH	Read, write, Heating control
10	003EH	Read, write, Motor control

The Register is 16 bit (2 bytes), HIGH in the front, and LOW at the back.
(3) Routine Data Manipulation
  • Function code 03: Read multiple registers.
    The starting address: 0000H~003EH, invalid if over range
    The length of the data: 0000H~0007H, Up to a maximum reading of 7 consecutive registers.
    The host sends: address + Function code + The starting address + Data length + CRC code.
    The response: address + Function code + Returns the number of bytes + multiple data of Register + CRC code.
  • Function code 10: Write multiple registers
    The starting address: 003AH~003EH, invalid if over range
    Register number: 0001~0004H, 4 registers will be the maximum for one continuous setting
    The host sends: address + Function code + The starting address + Write the register number + Number of bytes + Save the data + CRC code.
    The response: address + Function code + The starting address + the register number + CRC code.
  • Function code 05: Write one way switch output
    The host sends: address + Function code + carry-out bits + Data length + CRC code.
    The response: the same as the host sends in regard of the format and content
  • Handshake packet 4 bytes: 0X01 0X04 0X01 0XE3.
    Once the telecommunication line sets up, the master machine succeeds in connecting to the slave machine. After the communication status is on, no more handshake signal will be sent from the slave machine.
  • Error messages: If there is an error in the set parameters, return an error code, in order to debug and repair.

Format: Address code + function code + error code + CRC code, error message is as follow:

86: Incorrect function code. The received function code is not supported by the slave machine.

87: To read or write the wrong address of data. The designated data address is over the specified address range.

88: Illegal data values. The received data value from the host is beyond the scope of the corresponding address data.

2   Design of Perception Layer

2.1   Hardware Design of Sensor Nodes

Sensor nodes are able to collect the local data, control the field parameters and communicate with the master controller. Therefore, the hardware of sensor nodes can be divided into 4 parts: CPU processor, 485 communication module, sensor module and actuator module. The block diagram is shown in Fig. 2.

Fig. 2. Node composition block diagram

The quantity of processing data handled by nodes is much less than the master controller. Then, the nodes' CPU can use medium capacity series of single-chip like STM32F103. The STM32F103xx medium-density performance line family incorporates the high performance ARM® Cortex®-M3 32-bit RISC core operating at a 72 MHz frequency, high speed embedded memories (Flash memory up to 128 Kbytes and SRAM up to 20 Kbytes), and an extensive range of enhanced I/Os and peripherals connected to two APB buses. All devices offer two 12-bit ADCs, three general purpose 16-bit timers plus one PWM timer, as well as standard and advanced communication interfaces: up to two I2Cs and SPIs, three USARTs, an USB and a CAN [6].

485 communication module uses SP3485 chip, and the RO connects to the RXD of CPU(i.e. PA10 pin). DI connects to TXD of CPU(i.e. PA9 pin), while direction control signal DE & RE connect to PB0 pin, with AB as 485 bus. The module circuit is shown in Fig. 3.

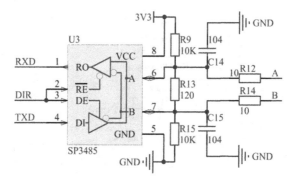

Fig. 3. Communication module circuit

Different sensor nodes have different interfaces. For better universality and instal-lation, the nodes have several interfaces: IIC, SPI, 1-Wire, UART, AD. Each interface has a dial-up switch for checking the fitness of its setting.

Actuator module mainly consists of multiple small relay and its corresponding drive. When it's running, if the power of the small relay's drive is insufficient, that relay will impel the big relay or contactor. In order to protect CPU and its circuit, the relay control will be isolated by optocoupler. The module driver circuit is shown in Fig. 4.

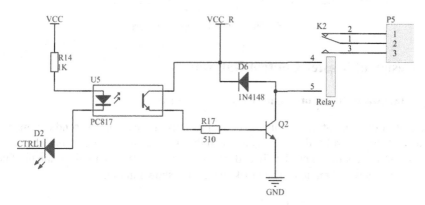

Fig. 4. Actuator module drive circuit

2.2 Software Design of Sensor Nodes

Sensor nodes take charge of three functions: reading sensor signal, controlling the relays due to the received signal and communicating with master controller. Before reading the signal from sensors, CPU will firstly scan the dial switch, read the sensor nodes and detect the sensor connected to the exact node so as to determine which algorithm should be adopted. Program flow chart is shown in Fig. 5.

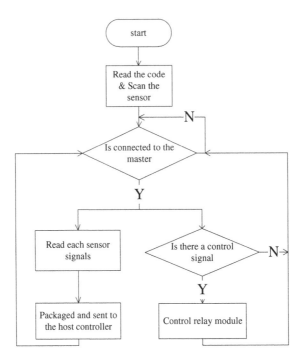

Fig. 5. The node program flow chart

3 Design of Master Control Gateway

3.1 Hardware Design of Master Control Gateway

The master control gateway is able to collect data from sensor nodes, send commands to the actuator control and communicate with the cloud server [7]. Therefore, the hardware of master controller can be divided into 3 parts: CPU, 485 communication module and internet network module. The block diagram is shown in Fig. 6.

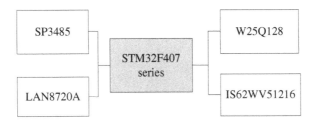

Fig. 6. The host controller composition block diagram

The master control gateway can handle complex task and require a speedy CPU, which is met by STM32F407. The STM32F407xx family is based on the high-performance ARM® Cortex™-M4 32-bit RISC core operating at a frequency of up to 168 MHz. The Cortex-M4 core features a Floating point unit (FPU) single precision which supports all ARM single precision data-processing instructions and data types. It also implements a full set of DSP instructions and a memory protection unit (MPU) which enhances application security [8].

To ensure the system can run smoothly, this system will use IS62WV51216 of 512 K × 16 LOW VOLTAGE, ULTRA LOW POWER CMOS STATIC RAM with RAM expanded and the flash expanded by W25Q128.

485 communication module uses SP3485 chip, and the RO connects to the RXD of CPU(i.e. PA3 pin). DI connects to TXD of CPU(i.e. PA2 pin), while the direction control signal DE & RE connect to PG8 pin, with AB as 485 bus and nodes connected by 485 bus. The module circuit is shown in Fig. 3.

Network communication module will use LAN8720A. The LAN8720A is a low-power 10BASE-T/100BASE-TX physical layer (PHY) transceiver with variable I/O voltage that is compliant with the IEEE 802.3-2005 standards [8]. The circuit chart is shown in Fig. 7.

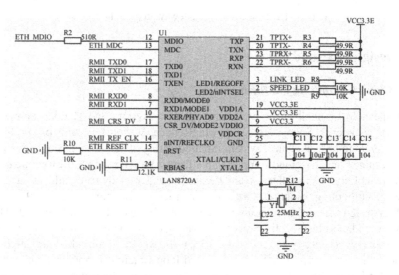

Fig. 7. The principle diagram of the network module

3.2 Software Design of Master Control Gateway

The master control gateway takes over three functions: collecting data from sensor nodes, sending commands to the actuator controller and communicate with cloud server. It can read the information from sensor nodes by roll poling [9], online update the list of sensor nodes regularly, collect the information from online sensor nodes regularly, encrypt the data and transmit it to the cloud server. When the command signal from cloud server is received, it will send out the command and make the nodes control actuators. Program flow chart is shown in Fig. 8.

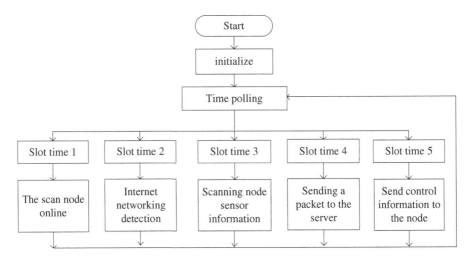

Fig. 8. The host controller gateway program flow chart

4   Design of Cloud Server

4.1   Cloud Computing Platform

With the help of flexible server, Linux operating system and Hadoop components (HBase, Zookeeper, Sqoop, Hive, Pig, MapReduce, Mahout), in view of two features of intelligent industrial control system: Small quantity of data for one single device and longer online time, this cloud computing platform is built through Infrastructure deployment technology of Virtualization and flexibility, under the dynamic allocation strategy to design computing resource, storage and internet. Please refer to the Fig. 9.

This design adopts the technology of Dynamic Feedback Load Balance [10], DFLB. Hadoop conducts cluster collection of the loading condition for current nodes and feedback to the scheduling system, which works as the weight of job scheduling algorithm. Through this way, a dynamic feedback closed-loop system is formed, which makes the cluster load gradually balance. Once all the nodes are of insufficient supply, this platform can apply to the system for more hardware resources (computing capacity, network bandwidth and storage space) in order to meet the application requirements.

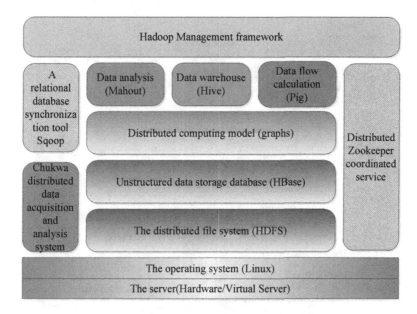

Fig. 9. Cloud computing platform architecture

4.2 Cloud Data Center for Designing Internal and External Network, Hybrid Encryption

According to the available data, if big data is not duly handled, user's privacy will be infringed badly, which is a big obstacle to promote smart industrial system. That's why the moment big data was raised its security issue attracted much attention. However, this system is designed within improved security policy, including configurable data acquisition platform for users and cloud data center storage.

For the user configurable data acquisition platform security strategy, according to the importance of the configurable data, three levels has been formulated: Level 1: running status of acquisition equipment and user log; Level 2: running status of acquisition equipment and user log (do not contain user address information); Level 3: only acquisition equipment running status data (do not contain user address information, time, etc.). The higher the level, the better user's privacy is protected. Users can configure freely in the intelligent master controller the gateway configuration page.

Cloud data center storage security [11] policy: firstly, to store data by independent internal and external network so as to realize logical isolation between the inside and the outside (basic data storing in the internal network, while the business data storing in the external network), which reduces network circuitry and equipment, make full use of the information available. By using advanced security means like firewall, user's information can be prevented from illegal invasion. Secondly, by using the password techniques, we can make sure the confidentiality and integrity of intelligent household data in cloud data center, as shown in Fig. 10.

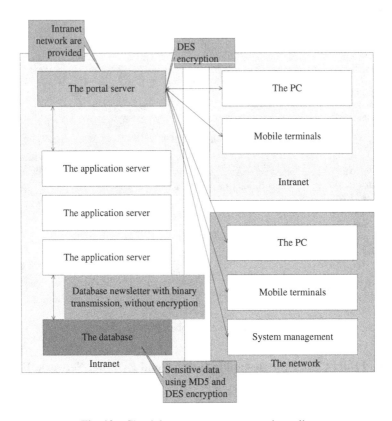

Fig. 10. Cloud data center storage security policy

Symmetric encryption algorithm (DES) can do the encryption speedily, but its short-coming lies in using the same key to encrypt & decrypt, which cannot guarantee the security or manage the key well; compared with asymmetric key system (RSA), although it's not necessary to undertake negotiation about key, a plan for public key management is needed. Moreover, asymmetric encryption algorithm is inefficient and slow, only suitable for encrypting a small amount of data. Intelligent industrial system has a large number of data source and the data quantity is big. It requires good timing, speedy encryption and efficient encryption algorithm. Thus, the symmetrical encryption algorithm is suitable for data encryption, however, the asymmetric encryption algorithm is suitable for the encryption of metadata or secret key.

5   Conclusion

With STM32F4 as the host CPU and STM32F1 as the nodes, this remote industrial control system also has its corresponding cloud service platform and data center framework, enable to achieve remote monitoring & controlling. Since its stability and data security are quite impressive, this system can realize automation of production system and optimal allocation of resources. This remote control system is suitable for manufacturing enterprises who intend to implement automation, to improve the quality of end product, to optimize the use of resources and to improve production efficiency.

Acknowledgments. The authors would like to thank Guangdong Province Special Project of Industry-University-Institute Cooperation (No. 2014B090904080), 2013 Guangdong Province University High-level Personnel Project (Project Name: Energy-saving building intelligent management system key technologies research and development) and the Project of Guangdong Mechanical & Electrical College (No. YJKJ2015-2) for their support in this research.

References

1. Lee, J., Kao, H.A., Yang, S.: Service innovation and smart analytics for industry 4.0 and big data environment. Procedia CIRP **16**, 3–8 (2014)
2. Liu, J., Wang, Q., Wan, J., Xiong, J., Zeng, B.: Towards key issues of disaster aid based on wireless body area networks. KSII Trans. Internet Inform. Syst. **7**(5), 1014–1035 (2013)
3. Want, R.: An introduction to RFID technology. IEEE Pervasive Comput. **5**(1), 25–33 (2006)
4. Liu, J., Wan, J., Wang, Q., Li, D., Qiao, Y., Cai, H.: A novel energy-saving one-sided synchronous two-way ranging algorithm for vehicular positioning. Mobile Netw. Appl. **20**(5), 661–672 (2015). ACM/Springer
5. Peng D, Zhang H, Yang L, et al.: Design and realization of modbus protocol based on embedded Linux system. In: International Conference on Embedded Software and Systems Symposia, pp. 275–280. IEEE press (2008)
6. Liu, M., Yu, J., Liang, H.: Wireless geotechnical engineering acquisition system based on STM32. Instrum. Techn. Sens. **5**, 95–97 (2010)
7. Liu, J., Wan, J., Wang, Q., Zeng, B., Fang, S.: A time-recordable cross-layer communication protocol for the positioning of vehicular cyber-physical systems. Future Gener. Comput. Syst. **56**, 438–448 (2016)
8. Gill, K., Yang, S.H., Yao, F., et al.: A zigbee-based home automation system. IEEE Trans. Consum. Electron. **55**(2), 422–430 (2009)
9. Liu, J., Wan, J., Wang, Q., Deng, P., Zhou, K., Qiao, Y.: A survey on position-based routing for vehicular ad hoc networks, telecommunication systems (2015). doi:10.1007/s11235-015-9979-7
10. Shu, Z., Wan, J., Zhang, D., Li, D.: Cloud-integrated cyber-physical systems for complex industrial applications. Mobile Netw. Appl. (2015). ACM/Springer. doi:10.1007/s11036-015-0664-6
11. Zhang, X., Du, H., Chen, J., et al.: Ensure data security in cloud storage. In: 2011 International Conference on Network Computing and Information Security (NCIS), vol. 1, pp. 284–287. IEEE press (2011)

Author Index

Printed in the United States
By Bookmasters